ADDITIONAL PRAISE FOR

JUNK SCIENCE

"A rare combination of erudition and moral outrage, this book should be read by citizens who worry about the future of the American republic, by educators who are failing to promote scientific literacy, and by scientists who bemoan the trashing of science."

— Jay Goldberg, professor, department of neurobiology, pharmacology, and physiology, University of Chicago

"Sends a strong, well-reasoned message of the increasing danger that this misuse of science presents to the ideals of American democratic society."

— E. Roy John, Ph.D., director, Brain Research Laboratories; professor of psychiatry, NYU School of Medicine

"Timely . . . not for the fainthearted. Intelligently forces us to confront the widespread misconceptions that are passed off as science and argues strongly that people and societies cannot hide behind a shield of ignorance any longer."

— Dr. Ian Turnbull, consultant neuroradiologist, department of neurosciences, Hope Hospital

JUNK
SCIENCE

An Overdue Indictment of Government,

Industry, and Faith Groups

That Twist Science for

Their Own Gain

DAN AGIN, PH.D.

Thomas Dunne Books

St. Martin's Griffin ☙ New York

THOMAS DUNNE BOOKS.
An imprint of St. Martin's Press.

www.thomasdunnebooks.com
www.stmartins.com

Design by Jamie Kerner-Scott

Library of Congress Cataloging-in-Publication Data

Agin, D. P.
    Junk science: an overdue indictment of government, industry, and faith groups that twist science for their own gain / Dan Agin.
        p. cm.
    Includes bibliographical references and index.
    ISBN-13: 978-0-312-37480-8
    ISBN-10: 0-312-37480-1
    1. Science—Moral and ethical aspects. 2. Fraud in science. 3. Pressure groups. I. Title.

Q175.35.A35 2006
500—dc22

                                                                    2006048115

First St. Martin's Griffin Edition: December 2007

10 9 8 7 6 5 4 3 2 1

To Joan, best friend and wife,
the rock of my journey

CONTENTS

PART SIX

GENES, BEHAVIOR, AND RACE

PART SEVEN

ON TRUTH AND LIES

JUNK
SCIENCE

INTRODUCTION

There is one overriding question that concerns us all: How can we get out of the fatal groove we are in, the one that is leading towards the brink?

—ALBERT SZENT-GYÖRGYI (1893–1986)

Yankee Doodle came to town, riding on a pony. Stuck a feather in his cap and called it macaroni.

—EDWARD BANGS, U.S. SONGWRITER

DURING THE NEXT THIRTY years the American public will suffer the consequences of a rampage against reason by special interests in government, commerce, and the "faith" industry. The public will pay a heavy price for the follies of people who think they can twist knowledge of the real world with impunity. And after thirty years, in the following generation, the problem may intensify. If current patterns of the corruption of reason remain in place, the twenty-first century will witness a descent into a dark era that will make the agonies of the twentieth century seem benign.

This book is about the corruption of science, in effect the corruption of reason, and the dangerous impact of such corruption on society.

Our subject is not "bad science" or weak science, but the socially destructive and often fatal twisting of science by special interests or by scientists themselves. I'll deal with a wide range of controversial issues in science and technology, but the focus is on corruption of science with societal effects that are often dangerous and sometimes catastrophic.

Our path can be changed. Every civilization has the choice of either controlling its own destiny or remaining passive as it awaits its future. What will we do? Our children and grandchildren will live through the consequences of our decision.

Since I'm a neuroscientist and biophysicist, and not a political or social scientist, my focus is more on science than on politics. Nevertheless, the politics is there between the lines, always there, since science is an archetypical human enterprise, and neither the enterprise nor its product can ever be separated from society.

Science and society are bonded by their mutual dependence and linked history, a bonding complex and often ignored by the public. Those in power are usually aware of the importance of science to the retention of power, but the public at large generally considers science in daily life a tedious abstraction—until serious consequences rivet its attention.

More than forty years ago, the historian Jacques Barzun said: "Very recently—within the last two or three years—the public decided that it wanted to know all about science. The publishers, docile creatures, have therefore brought out books by the barrelful."

The sudden interest of the American public in science in the late 1950s was due primarily to the public realization that one of the products of science, nuclear weapons, in the context of the Cold War with the Soviet Union, might very well result in the obliteration of civilization. Science and books about science were suddenly important. Unfortunately, the public interest in science never translated into public scientific literacy—a knowledge of fundamentals—adequate to defend the public against junk science—against junk science promulgated by the faith industry, against dangerous government junk-science policy, dangerous corporate manipulations of science, and hazardous junk science in the health-care marketplace.

And no matter the barrelful of books, scientific literacy has also remained inadequate for the understanding of nuclear weapons and nuclear warfare. Although the so-called Star Wars strategy of defense against nuclear weapons has been rejected by most physicists as an unworkable scheme and a waste of taxpayer money and resources (see Chapter 12), the public, ignorant of scientific realities, would rather believe Star Wars politicians who have no more understanding of the physics of nuclear warfare than the public itself.

The current situation is not good. It's not good when science and technology, the two prime generators of political, social, and economic progress, are not understood, misunderstood, or abused in ways dangerous to society.

❀   ❀   ❀

MANY PEOPLE BELIEVE WE'RE living in an "age of science." We're certainly living in an age of scientific discovery, but given the lack of acceptance of science by the public, that this is an "age of science" is questionable. The public uses science but doesn't fully accept it, a paradox of dangerous consequence.

Dangerous? Immensely so. Ignorance of science and denial of realities revealed by science are responsible for more human misery and death each year than any other factor acting in human affairs, from wars to floods to earthquakes.

"Ignorance of science" refers to the gap between public knowledge and the knowledge of scientists—a gap for the most part man-made.

And what is science? Working scientists don't spend much time concerned about definitions of what they do. For most working scientists, science is essentially an attitude, an insistence that conclusions about the natural world must be objective and drawn only from evidence. Everything else is speculation at best or superstitious myth at worst. The so-called scientific method is really a collection of methods of observation that can be used to lead to such objective conclusions.

In understanding the nature and impact of science in the modern world, it's important to realize that science and technology cannot be separated. The reason is simple: Basic science produces new applied science (technology), and new technology in turn produces new methods of doing basic science, which leads to new science that produces new technology, and so on. The science-technology spiral is unending, and because of this spiral it's not possible to predict with any certainty where science and technology will go in the future. This is an old idea that as a student I first heard articulated by the philosopher Karl Popper (1902–1994), but no doubt there are many versions of it. The idea is significant because it means, among other things, that we need to be careful about judging a particular research path in science or technology as "unimportant." No one, not scientists, policymakers, or politicians, can be certain about such judgments. The direction of science and technology and the impacts of science and technology on society are in the long term fundamentally unpredictable. Which means potential dangers or benefits of a particular path are also unpredictable. Commentators may moan (and be paid handsomely for their moaning) about the potential dangers of science and technology, but the usual case is that they don't know: The dangers are not clear, and in fact the bemoaned path of science or technology will often produce unpredictable beneficial results we could not even imagine beforehand.

GIVEN THE ABOVE DEFINITION of science, what is "junk science"? I define junk science as extensively corrupted science, science corrupted in objectivity and/ or method, the corruption either deliberate or involving sloppy methods or due to ignorance of what science is about, the outcome useless conclusions that make false statements about the natural world. Proper science can also make false statements, but in proper ("real") science there is no extensive corruption of objectivity or methods of observation. In proper science, false conclusions may occur because of insufficient evidence or because of flawed interpretations of evidence, but this is the ordinary operation of science, and one of the main agendas of proper science is the continual correction of false conclusions.

And what is "bad science"? How do bad science and junk science differ? I use the term "bad science" to refer to science whose procedures in an isolated case are flawed, for example, a lack of adequate controls, or an instance where an interpretation is based on false assumptions.

In general it's a question of whether one is talking about a single experiment or a whole field of investigation. When a whole field is based on bad science and/or twisted science and/or fabrication, we call that field "junk science." So it follows that while creationism is junk science, a single experiment on the origin of life that involves inadequate controls is only bad science.

Often the most important characteristic of a junk science is that much of it is based on no science at all, but on wishful thinking, fantastical thinking, and so on.

But even if we can define some terms, we need to recognize that the complex interface between science and politics does not always lend itself to easy categorizations. In general, any political group or movement that twists science with bias in order to support a particular agenda is producing junk science, even if the twisting is subtle rather than blatant. Twisting with bias is a corruption of science, and when the twisting is extensive and coupled to an agenda, we have junk science in a possibly dangerous mode.

One form of twisting is deliberate fabrication. Wholesale fakery of an entire repertoire of experiments produces a junk science. I treat fakery in some detail because I think it's important to understand what it is and how and why it exists.

And it's also important to be aware that science is a human enterprise whose people are subject to all the vulnerabilities of people in other enterprises. This book, after all, rides on an undercurrent of the "sociology" of science, an undercurrent in a sea of complexities.

Junk science is often a deliberate tool used by those who seek to exploit the public for one end or another. Most Americans don't read much, which means the majority of Americans get their science information primarily from television news and entertainment shows. They're the prime market of corporations, politicians, hucksters, and some consumer activist groups who seek profit and power by manipulating a huge public sector ignorant of scientific realities.

Junk science is often the handmaiden of dogma. The history of dogma versus science, from Church persecution of researchers like Galileo, to Nazi racist terror, to the imbecilities of Soviet social and biological research, is a testament to how easily whole societies can be maneuvered like blind sheep for political ends when such societies know only junk science.

In general, science is the human examination of reality. Junk science is the clouding of that examination by carelessness, fear, myth, and deliberate falsehood.

WE'RE NOT CONSIDERING ALL of junk science in this book. Our focus is junk science that has societal effects that are both significant and dangerous. In the chapters that follow, we'll see that junk science, the corruption of science, is currently more pervasive and more dangerous than most people imagine.

The first chapter deals with junk science in the service of dogma, from Galileo in the seventeenth century to the Soviet Union in the twentieth century, three hundred years of the corruption of reason with catastrophic societal consequences.

All of the second chapter is devoted to fraud (fabrication) in science, with examples of scientific fraud that had significant impact on either science or society or both, and a discussion of the current prevalence of fraud in the scientific enterprise.

The following sixteen chapters examine examples of modern junk science, providing a detailed discussion of junk science impacting current life in a significant way.

The final chapter discusses the failures in our current society that have allowed dangerous junk science to exist, and often thrive, in the public arena.

❀   ❀   ❀

No one can write a book like this one alone, since the author depends too much on the work of other people, on their ideas and on their discoveries, and on people who help directly or indirectly. I thank the many fellow scientists who over the years have caused me to maintain my faith in the human enterprise we call science. Many of them have never been aware of their contribution. I also thank the librarians of the University of Chicago and the Chicago Public Library. My editor, John Parsley, deserves kudos for his hard work. And special thanks go to my agent, Carolyn French of the Fifi Oscard Agency, whose sure hand guided this book to publication. And finally I thank the readers of *ScienceWeek* for showing me during the past ten years how committed many people can be to knowing the truth about the world.

PART ONE

THE MOONS OF JUPITER

1

SCIENCE, JUNK SCIENCE, AND DOGMA

Rome has spoken; the case is concluded.
— St. Augustine of Hippo (354–430)

There is no place for dogma in science. The scientist is free,
and must be free to ask any question, to doubt any assertion,
to seek for any evidence, to correct any error.
— J. Robert Oppenheimer (1904–1967)

WHAT HAPPENS WHEN DOGMA rules? We need the past as a guide, since only through the lens of history are the realities of the behavior of whole societies completely visible. To ignore our past, to avoid learning from it, is a mindless attitude that only increases the likelihood of personal and social calamity. What a tragedy it is that history is hardly ever taught to children as a cautionary tale. Instead, history is taught as an exercise in tribal self-glorification, a pedagogical scheme that may be useful to politicians in their manipulations of the public, but a scheme that in the long run produces social dangers and the catastrophes of war.

GALILEO AND THE MOONS OF JUPITER

So we begin with Galileo Galilei (1564–1642), a man whose very name has come to signify the perpetual battle between dogma and science, a battle won by dogma, a defeat now recognized as a disaster for human society.

The story of the Church against Galileo has been repeated (and often distorted) over and over again in history and literature. But what was the crux of it? Some say that the officials of the Church of that time were aware of the truth of Galileo's assertions that the Earth revolved around the Sun, but were incapable of publicly admitting this because of fear of demolishing the philosophical structure upon which the Church rested—the theological position, originating with the ancient Greeks, that a mechanistic interpretation of nature could never be more than a model, an intellectual artifact, since between theory and reality there would always be a gap that could not be bridged by human reason. The Church had received from the ancients a fundamental view of the cosmos that the Church had preached since the beginning of Christianity, and that view could not be denied without demolishing the foundations of the religion itself. At least, according to this interpretation of the crux of the conflict, that was the view of Church officials of the seventeenth century. Of course, eventually, after two hundred years, the Church did accept the Galilean/Copernican view of the solar system, and without destruction of its theological foundations. (Some may argue that if anything the foundations were strengthened.)

The other view of the crux of the matter is simpler and focuses on the elemental battle between dogma and reality, the refusal of the dogmatists to acknowledge reality, the stubborn efforts of the dogmatists to contrive and deny even when one is handed a telescope and told to look at the moons of Jupiter and see whether or not they are real. So goes the story of the Church and Jupiter's moons, although if officials of the Church refused to look, many academics, the so-called philosophers of Pisa, also refused to look.

Why not look? Because to look and see what Galileo (and others) said could be seen would demolish the foundations of one's reality. The dogma was that the Earth did not move. And even of those who accepted the Copernican idea that the planets (other than Earth) revolved around the Sun, many would not accept the idea that the Earth itself revolved around the sun—because they believed the Earth would then lose its moon. Thus, to see the moons of Jupiter was to understand that a planet could revolve around the sun without losing its moons, and that the Earth could do this also.

Here are Galileo's own words about the import of Jupiter's moons:

> Here [in the Jovian moons] we have a powerful and elegant argument to
> quiet the doubts of those who, while accepting without difficulty that the
> planets revolve around the Sun in the Copernican system, are so disturbed to
> have the Moon alone revolve around the Earth while accompanying it in an

annual revolution about the Sun, that they believe that this structure of the Universe should be rejected as impossible. But now we have not just one planet revolving around another while both make a large circle around the Sun, but our eyes show us four stars that wander around Jupiter, as does the Moon around the Earth, and these stars together with Jupiter describe a large circle around the Sun in a period of twelve years.

But the hard evidence that there were indeed people who refused to look at Jupiter's moons is scanty, most of the evidence in comments by Galileo himself. The best surmise is that there were indeed people, academics, philosophers, Church officials, who refused to look, even when others did look and were looking all over Europe as soon as the announcement of the Jovian moons was made. And if they looked, did they believe the moons were there? In this context, the important point is not who looked and who refused to look—no, the important fact is that there were probably enough people of substance, even of eminence, people of the established order who refused to look, assuming that Galileo did not concoct the idea of refusals, as some have suggested.

Galileo discovered the moons of Jupiter in the year 1610. On June 22, 1633, he received the final sentence of the Church, with the following words read out to him:

You have rendered yourself vehemently suspect of heresy, namely of having held and believed a doctrine which is false and contrary to the Sacred and Divine Scriptures, that the Sun is the center of the world and does not move from east to west, and that the Earth moves and is not the center of the world; and that one may hold and defend as probable an opinion after it has been declared and defined contrary to Holy Scripture.

Never mind Galileo's subsequent recantation, the question of who looked or who did not look, the question of how many Church officials quietly accepted the reality of Jupiter's moons, the crux of the matter, the essence of dogma, the fundamental and unresolvable confrontation between dogma and science is clear in the above paragraph, in the accusation read to scientist Galileo Galilei as on Wednesday, the 22nd of June in the year 1633, he knelt on the floor in a room adjoining the church of Santa Maria sopra Minerva in Rome. His crime was to refute a doctrine with a telescope, counter a dogma that suddenly, with the invention of the telescope, became a dogma based on junk science.

And the junk science endured. When Harvard University was founded in the year 1636, the assembled university scholars did not accept Galileo's work and they remained firmly committed to the Ptolemaic theory of the universe. Were they too busy to look at Jupiter's moons?

Galileo's major work on the solar system, *Dialogue Concerning the Two Chief World Systems*, was not removed from the Roman Catholic index of prohibited books until 1835, two hundred years after the Church forced his recantation.

Dogma is not easily melted.

PHRENOLOGY AND INHERITED TRAITS

THE JUNK SCIENCE FOISTED ON human society by the Church during the two centuries following the invention of the telescope concerned man's view of the world. At about the same time the Church finally accepted the reality of the Galilean world, a new junk science arose concerning man's view of himself, his view of his mind and brain, and this new junk science came to be a precursor of twentieth-century catastrophe.

Both science and junk science have a powerful impact on the public when concerned with mind and brain, the workings of intellect and emotions, and the differences in individual character.

In the eighteenth century, knowledge of human physiology—and of the workings of the brain and nervous system—was primitive, sometimes fanciful, and often overshadowed by centuries-old myths and religious doctrine. There was still uncertainty about whether the brain was the organ of mind, uncertainty about the function of nerves. Muscle action was a mystery, the connections between nerves and muscles not known until the 1870s, and electricity was for the most part a parlor trick used to entertain the aristocracy in their salons. But no matter what Aristotle or the Church or Descartes decreed, many physicians understood, on the basis of repeated clinical observations of traumatic injuries, that the brain had a great deal to do with mental faculties, both ordinary and peculiar. And if mental faculties could be rationally categorized, the faculties given this or that name and differentiated one from the other by observation, was it not possible that the brain itself was organized in a similar fashion?

The idea that the anatomical organization of the brain is related to its various functions is more than two hundred years old and is today called "localization of brain function." The existence of considerable localization of function is indisputable: There are brain regions involved with specific primary inputs such as vision, audition, taste, etc.; brain regions for specific primary outputs to various muscle systems; and brain regions for speech and the understanding of language. The still unclear aspects concern anatomical localization of other so-called higher faculties, e.g., learning, memory, perceptual analysis, motivations, and various other cognitive abilities. But the general idea of localization of brain function is not new; it arrived in Europe two centuries ago—and rapidly degenerated into junk science.

At the end of the eighteenth century, the neurologist Franz Joseph Gall (1758–1828) proposed a view of the brain and mental function that quickly swept both sides of the Atlantic, flowered in public lecture halls and magazines, and drew the approval of people in upper-class drawing rooms because of the tenet that mental faculties were fixed from birth by inheritance and incapable of modification by education. European and American aristocracies adored (and still adore) the fiction of genetic inferiority and ineducability of the lower classes, since it excuses their privilege as a natural consequence of good breeding (and helps to justify the treatment of the servant class as half-wit children). Without the support of the upper classes, it's doubtful that Gall's modest neurologic research would have exploded into the almost frenzied popular fad of the early nineteenth century.

Gall's idea, that the brain was organized into twenty-seven "organs" (regions of specific function), and that these structures corresponded to particular protuberances on the external cranial surface, was simple in conception but revolutionary in practice, since it meant that with a proper "map" in hand, the contours of anyone's skull could be examined and then statements made about that person's mental faculties. These faculties were presumed to be innate and incapable of modification.

Gall's pupil, Johann Casper Spurzheim (1776–1832), lectured extensively in the United States, promoting an American phrenology craze as intense as that in Europe. The popular rage was to have one's head "read" by a phrenology specialist—unless one happened to have bumps in the skull around the ears, since those bumps were associated with combativeness, destructiveness, secretiveness, acquisitiveness, and a devotion to food. The junk-science frenzy eventually reached a point where some criminologists claimed to be able to detect a murderer by the shape of his or her head.

Thus, a sensible and valuable idea, that of the localization of brain function, coupled with fallacious ideas concerning inherited mental faculties and the significance of conformations of the skull, became mangled into a junk science that in the nineteenth century deceived several generations of Europeans and Americans. All for the love of the idea of inherited mental faculties, an insidious dogma that continued into the twentieth century and set the stage for catastrophe.

As for phrenology itself, its principles were demolished in 1870 when experiments produced the first functional maps of the cerebral cortex, maps totally unrelated to any contours of the skull. Of course, to some people experimental science was irrelevant, and phrenology demonstrations continued as a drawing room entertainment well into the Edwardian era at the turn of the century.

EUGENICS IN AMERICA

RECENT MAJOR ADVANCES IN MOLECULAR genetics and genetics biotechnology have produced in some quarters a "genetics euphoria," with books in the marketplace called *Your Genetic Destiny*, with many researchers rushing to establish the "genetic basis" of various human "behavior traits," the latter term encompassing, by implication, both "good" and "bad" behavior traits, and of course reminiscent of the phrenology craze in the nineteenth century. Can we expect a future proposed genetics program to eliminate what proponents might call genetically based "bad behavior traits"?

We have already traveled that miserable road. One of the most disastrous examples of the ignoble application of science is that of the so-called science of "eugenics" in the first half of the twentieth century, a junk science that directly spawned genocide.

One might think that the United States, the so-called melting pot of nationalities, ethnicities, cultures, and religions would be difficult ground for the cultivation of the idea that some groups are better than others, more "fit," and that the least fit ought to be outbred (or even sterilized) as quickly as possible to prevent contamination of the national gene pool. But the facts are otherwise, since the eugenics movement had considerable strength in the United States well past the middle of the twentieth century, supported by a number of prominent intellectuals and scientists, and of course supported by many others who saw themselves as members of the "fittest."

The term "eugenics" was apparently coined in 1883 by the British statistician Francis Galton (1822–1911), a first cousin of Charles Darwin (1809–1882). To Galton, the term meant "well-born" and referred to an effort to encourage the "best" people in society to have more children (positive eugenics) and to discourage or prevent the "worst" elements of society from having many, if any, children (negative eugenics). Eugenics became solidified into a movement in various countries throughout the world in the first three decades of the twentieth century, but nowhere more solidly than in the United States and Germany.

During the first thirty years of the twentieth century, eugenicists attempted to analyze the inheritance of various diseases and behavioral traits in studies of correlations between relatives and studies of family pedigree charts. The basic assumption was that if a trait recurred in families over several generations it must be genetic. The American eugenicist Charles B. Davenport, director of the Station for Experimental Evolution and the Eugenics Record Office at Cold Spring Harbor, Long Island, New York, constructed elaborate pedigrees for Huntington's chorea, albinism, epilepsy, feeblemindedness, and thalassophilia or "love of the sea" (which eugenicist Davenport proposed to be a Mendelian sex-linked recessive trait especially prominent in the families of naval officers). Harry H. Laughlin, superintendent of the Eugenics Record Office (a federal agency), studied the inheritance of criminality, feeblemindedness, and many other deleterious traits in different ethnic and racial groups. Eugenicist Laughlin concluded that Eastern Europeans, Mediterraneans, and Russian Jews, among others, harbored a large number of defective genes in their populations. Such studies, sprinkled with anecdotes, formed the backbone of eugenics "science," a junk science that deceived the American public with the tacit approval of many American scientists.

American eugenicists also worked to establish eugenics-based legislation in the United States. Eugenicist Laughlin was appointed "Expert Eugenics Witness" to the House Committee on Immigration and Naturalization in 1921. His prison and hospital data were critical in convincing the committee that America's germplasm was being weakened by mixing with the lower quality genes coming from Southern and Eastern Europe, the Balkans, and Russia. This led to passage of the Johnson-Reed Act in 1924, which restricted immigration from these regions. Laughlin and others also lobbied at the state level for the passage of eugenic sterilization laws, which would allow individuals in state institutions to be forcibly sterilized if they were judged to be genetically defective. More than thirty-five states passed and used such laws, and by the 1960s, when most of these laws were beginning to be repealed, more than 60,000 people in the United States had been sterilized for eugenic purposes. In Germany, the National Socialists (Nazis) used Laughlin's model as one of the bases of their sweeping sterilization law of 1933, which ultimately led to the sterilization of over 400,000 people.

Clearly, anyone interested in an accurate history of twentieth-century Nazi Germany needs to be aware that the Nazis clipped their eugenics sterilization laws directly out of the American agenda.

The American eugenics movement did not wither away easily. As late as the 1960s, noted biologist and endocrinologist Dwight J. Ingle (1907–1978), member of the U.S. National Academy of Sciences and chairman of the Department of Physiology at the University of Chicago, founder and long-time editor of the influential journal *Perspectives in Biology and Medicine*, continued, in widely read published articles, to call for the mass eugenic sterilization of American blacks to prevent "weakening" of the U.S. Caucasian gene pool.

A similar public call for mass eugenic sterilization of blacks was made during that time and later by the noted physicist and engineer and Nobel laureate William B. Shockley (1910–1989), also a member of the U.S. National Academy of Sciences. Memoirs concerning Shockley in current science publications rarely mention his sociopolitical activities in the 1960s and later.

All of which brings to mind a statement by the biochemist Erwin Chargaff, published in of all places the journal *Perspectives in Biology and Medicine* (1973): "Outside his own ever-narrowing field of specialization, a scientist is a layman. What members of an academy of science have in common is a certain form of semiparasitic living."

I second Chargaff's view. I worked in the same academic department as Dwight Ingle for more than a decade. I knew him as a soft-spoken, congenial, charming fellow, a crack endocrinologist in his time, but a total novice in the fields of psychology, statistics, and genetics, the fields that supplied (or ought to have supplied) the fundamentals for his socially destructive, crazy, and oft-repeated junk-science warnings that America should sterilize its blacks in order to protect the American "gene pool."

Yes, Chargaff is right: Outside their specialty, scientists are laymen—and sometimes a great danger to the very public that floats their eminence.

EUGENICS IN NAZI GERMANY

THE EUGENICS MOVEMENT IN THE United States ultimately failed, maybe because the Great Depression of the 1930s had the public focused on daily economic survival rather than on cleansing the American gene pool, or maybe because the man in the White House, Franklin Delano Roosevelt (1882–1945), was either too knowledgable or too astute a politician to promote eugenics as government policy. Certainly the American people, usually forward-looking and optimistic, have a natural cultural resistance to any view that holds that their individual destinies are already foretold in their genes. In the United States, eugenics, despite its influential supporters, was eventually devalued by the federal government. In contrast, in Nazi Germany, at the same time, eugenics became a priority government program that quickly transformed into a literal reign of terror.

Sixty years after the end of the Nazi regime in Germany (1933–1945), studies of the active collaboration of a number of German scientists with the Nazis continue to be a focus of attention. Maybe part of the reason for the attention is puzzlement: These scientists actively collaborated with a tyrannical regime whose essence was totally opposed to the very spirit of science. Hitler, in fact, is said to have dismissed German physics with a wave of his hand and a statement that Germany could do without physics for a thousand years. What was in the minds of these scientists when they chose to actively support the Nazis? Was it an arrogant belief that their expertise in a science gave them superior insights into the enigmas of political, social, and economic realities? Historians, sociologists, and psychologists will continue to ponder such questions.

Meanwhile, the contemporary German science community is struggling to deal with its past, a past that's a sad story in the history of science. We now know that in Nazi Germany, physicians and medically trained academics, many of whom were proponents of "racial hygiene," or eugenics, legitimized and helped to implement Nazi policies that aimed to "cleanse" German society of people viewed as biologic threats to the nation's health. Racial-hygiene measures began with the mass sterilization of the "genetically diseased" and ended with the near-annihilation of European Jews. Anti-Semitism became a scourge in the German scientific community. Physicians and biologists who supported eugenics had to accommodate themselves to Nazism's rabid anti-Semitism. Many medical and biological scientists accepted the persecution of Jews because they favored the new emphasis on biology and heredity, the increased research funding, and the new career opportunities — including openings created by the purge of Jews and leftists from the medical and public health fields.

Ernst Rüdin, director of the Kaiser Wilhelm Institute for Psychiatry in Munich, and Ernst Fischer and Otmar von Verschuer, both of whom headed the Kaiser Wilhelm Institute for Anthropology in Berlin during the Nazi era, advised the Nazis at the highest levels. All three men had direct contact with the Nazi leadership and served on important government advisory panels. Rüdin sat on an expert committee to the ministry of the interior on population and race policies. There is evidence that Rüdin, whose work to give racial laws a scientific basis was funded directly from Hitler's office, chaired the committee's working group on "racial hygiene and racial policies." This panel set the criteria for the castration of criminals and the forced sterilization of so-called inferior women, particularly those with "psychological" problems.

According to a recent report, Rüdin lobbied successfully for ever broader criteria, and on Rüdin's initiative, the sterilization came to include the "morally ill" — the Nazi term for the mentally handicapped. This category covered 95 percent of the 400,000 sterilizations carried out between 1933 and 1945. At Rüdin's suggestion, the sterilized included 600 children born of black French soldiers and German women in the state of Rhineland, which the French occupied after World War I. Perhaps the most significant aspect of all of this is that these policies, which now seem the product of deranged minds, were not proposed and implemented by a few mentally unbalanced political leaders, but were indeed proposed and implemented by at least part of the German scientific establishment. Why did this happen? And how can the present scientific community prevent such a thing happening again?

The nineteenth-century junk-science myths concerning inherited human qualities were carried into the twentieth century as eugenics and soon became instruments of violent social destruction. Millions died because of fallacious ideas concerning biological determinism.

But was this a failure peculiar to the German science community? The evidence suggests otherwise: Had the United States been a dictatorship during the 1920s and 1930s, the programs advocated by American eugenicists might have—indeed, probably would have—been implemented by an American government that had already incorporated many of the ideas of eugenics into government policies. There is not, and never has been, much difference between German scientists and American scientists as scientists.

Eugenics, a strong movement in both Germany and the United States, came to different ends in these countries because the two countries had different political histories in the years following World War I—and not because of "national character" differences in the science communities.

LYSENKOISM IN SOVIET RUSSIA

IF GERMAN SCIENTISTS AND AMERICAN scientists were not much different as scientists, the same is true of Soviet scientists between 1917 and the end of the Soviet regime. The impact of politics on science becomes clear when one considers that at the same time (in the 1920s and 1930s) that the American and German science communities were supporting or promoting junk-science eugenics as a movement with a scientific basis, Soviet scientists were supporting or promoting Marxist-Leninist junk science that proposed a view completely opposite to that of biological determinism—the view that hardly anything is biologically determined, that all individual differences are produced by circumstances and environment following birth and early development.

The individual known in the West as the prime example of the corruption of science in the Soviet Union is Trofim Denisovich Lysenko (1898–1976). Born in the Ukraine, Lysenko was an agronomist who became interested in the cultivation of new varieties of plant forms, and before long he began proposing that acquired characteristics of plants could be inherited, a view in sharp contradiction to the findings of geneticists around the world, particularly those in Western Europe and the United States.

This view (for both plants and animals) was not at all new, having been proposed more than a hundred years before by the French biologist Jean-Baptiste Lamarck (1744–1829), the first to formulate in detail a theory of evolution, the idea of gradual modification through time. Lamarckianism included the following basic ideas:

1. The similarity between many species is accounted for by descent from a common ancestor.

2. The origin of variations among members of a species are caused by the environment.

3. Particular organs and structures evolve through time because use enhances the development of adaptive variations and disuse eliminates nonadaptive variations.

4. The variation between species is due to each species responding to different environmental needs by developing new organs or discarding old ones.

Of the above ideas of Lamarckianism, only the first is accepted by modern biology; all the others have been discarded during the past two hundred years of biological research.

Lamarckianism is, in effect, "environmentalism," the idea that environment is the prime factor determining individual and species destinies. This view is in complete agreement with the views of Marx and Engels, who developed the "scientific" aspects of Marxism in the early part of the nineteenth century, during the period when Lamarckianism was a well-known theory in biology. When the Soviets came to power in 1917, and especially after Lenin's death in 1924, the Marxist view of evolution became Soviet doctrine, and with it the essentials of obsolete Lamarckianism.

But obsolete or not obsolete, Lamarck's ideas were adopted by Lysenko. One reason may have been Lysenko's ignorance, since apparently he hardly read any scientific literature at all. Despite what some have suggested, Lysenko was not put up to his views by the Soviet government. Instead he formulated his views of inheritance of acquired characteristics in plants first, and he was then "discovered" by the Soviet regime and used for their purposes. Lysenko reported that he could alter the genetic constitution of strains of wheat by appropriate control of the environment, an idea contradicted by an already huge body of research in Western countries. In this case, at least, the Soviets did not manufacture junk biological science, they had the fallacious Marxist approach to evolution as a start, and Lysenko provided them with fabricated evidence to support the Marxist theory.

Lysenko's ideas mainly involved agriculture. The Soviet Union suffered serious wheat shortages during the 1930s, and Lysenko promoted the idea of improving wheat crops by changing wheat genetics through exposure to special environments. Later he claimed that wheat plants raised in appropriate environments could produce seeds of rye, a completely different species of plant. It was all fakery and ballyhoo promoted by the Soviet government and derided by the world's agronomists.

With a proper proletarian background, and views that supported Soviet pseudoscience, Lysenko was touted as a hero by the Soviet press, and he quickly rose to be a dominating figure in Soviet biology. He pronounced the Mendelian theory of heredity to be wrong, and as director of the Institute of Genetics of the Soviet Academy of Sciences from 1940 to 1965, Lysenko was ruthless in silencing all scientists who opposed him. Dissenting biologists "disappeared" or were put on trial and sent to gulags.

That is Lysenko's story, but it's not the story of all of Soviet science, not even of all of Soviet biology. There were some Soviet biologists who understood the intellectual corruption that had been set in motion by the Soviet government. Ivan Pavlov (1849–1936), the biologist who elucidated the conditioned reflex, and who at the time was the most famous biologist in the Soviet Union, was not only an anti-Communist, but vociferously so. In 1934, at the age of eighty-five, Pavlov wrote to the USSR Council of People's Commissars and said: "You disperse not revolution, but fascism, and with great success throughout the world. . . . You are terror and violence."

But the damage was done, and people like Pavlov could not save Soviet biology. Soviet physics and engineering, intrinsically useful to the Soviet regime in military matters and in its effort to industrialize the USSR, thrived under the Soviets—as long as the physicists and engineers avoided murder or banishment by refraining from anti-Soviet commentaries. Soviet biology, in contrast, was completely wrecked by Lysenko and his patron Stalin. Scientists throughout the world considered publications by Soviet biologists untrustworthy, and until the 1990s hardly any scientific paper published by a Soviet biologist was received as serious work in the West. Soviet biologists were found to be too ignorant of the world scientific literature, and in the case of geneticists, too prone to fabricate data in order to be in agreement with Lysenkoism.

The social agonies of Nazi Germany and the Soviet Union are now history, but Nazi science and Soviet science will stand for centuries as prime examples of nefarious government corruption of science and the science community.

2

FRAUD AND FABRICATION IN SCIENCE

The worst crime is faking it.
— KURT COBAIN (1967–1994)

Why do we think that the devotees of Newton's laws will be
more saintly than those ruled by Cardinal Law?
— RICHARD LEWONTIN

ALBERT **E**INSTEIN **(1879–1955) OFTEN** spoke of the "temple" of science, a place of devotion to the search for truth. And if such a temple exists, then faking evidence is certainly a desecration of that temple — and from the inside.

Fakery in science is something special. Junk science that involves corruption by dogma and junk science that involves corruption by fraud are not the same. On occasion the dogmatist will engage in fakery, but the usual dogmatist scenario involves deliberate selection of valid scientific results or total abandonment of science to support a political or social agenda. In contrast, junk-science fabrication is deliberate faking of data, sometimes to support a political or social agenda, but usually to produce personal gain for the faking scientist or for a small number of associated scientists. Faking data is so abhorrent to the average scientist that most scientists consider themselves personally violated when they learn of an instance, even if the fabrication is in another field. Scientists are not monks, but most of them do have a commitment, and desecration of the temple by one of their own is never received without pain.

The cases described in what follows are not trivial in either the degree of fakery or in the consequences for science and society. In the Abderhalden case, for example, the consequences were literally deadly for large numbers of people.

It's a common assumption that science is automatically self-correcting, or even that scientists are somehow a nobler breed above the failings of the remainder of humanity. But these ideas, and the idea that fakery in science is rare, are myths. In this age when science and technology have maximum impact on society, the public, for its own protection, needs an understanding of the fundamentals of the scientific enterprise. It's not true, for example, that any scientific paper published in a reputable scientific journal is automatically a reliable source of evidence.

How does serious fakery originate and how does it unfold in science? The modern era has given us several cases of scientific fakery whose details are known well enough to reveal a few things about how (and maybe why) it happens.

THE PILTDOWN HOAX

AT THE BEGINNING OF THE twentieth century, several interrelated intellectual debates about human origins were ongoing in anthropology. An important debate focused on the order in which the major anatomical changes occurred in human evolution. One idea was that the first step on the evolutionary path to humanity was the adoption of upright locomotion (bipedalism). A second view held that development of a large brain was the key event, such development producing an intelligent but still arboreal animal. Another debate concerned geographic origins: Did the first humans emerge in Eurasia or in Africa? With scanty fossil or other physical evidence to support any particular theory, controversies persisted and became popular items in the press.

The British anthropologists of that time held a strong belief in the Eurasian origins of early man and in the importance of the development of a large brain, the organ that differentiated man from other animals. British anthropologists, however, were at a working disadvantage: No important human fossils had been discovered in Britain. France and Germany had important digs, but nothing existed in the British Isles. Then, in the years 1910 to 1912, as if a gift from some anthropology muse with a liking for the British, a new fossil find dropped into the lap of British anthropology and immediately caused a sensation.

The main actor in what would become one of the most notorious scientific fiascoes of the twentieth century was an English lawyer and antiquary named Charles Dawson (1864–1916). An avid amateur fossil collector, Dawson (like all fossil collectors) kept his eye on workmen's diggings associated with new construction or repairs. As early as 1908 (the exact date is not known), he was apparently strolling down a country road near Piltdown Common in Sussex when he noticed some strange gravels, made enquiries, and soon after entered the source of the gravels, a gravel pit on a nearby farm. After several return visits, Dawson found what appeared to be skull fragments of unusual thickness. He eventually brought the fragments to Arthur Smith Woodward (1864–1944), the keeper of paleontology at the British Museum of Natural History in London, and Woodward and Dawson immediately began serious excavations at the Piltdown site. With the help of the French cleric turned paleontologist Pierre Teilhard de Chardin (1881–1955), between 1910 and 1912 they assembled several contiguous occiput-temporal skull fragments comprising about one-fourth of a cranium, and part of a lower jaw—the group of fragments considered by them to be the remains of a single individual.

News of the discoveries reached the press, and on November 12, 1912, *The Manchester Guardian* headlined the story: THE EARLIEST MAN? A SKULL MILLIONS OF YEARS OLD. ONE OF THE MOST IMPORTANT OF OUR TIME.

The article proclaimed that the Piltdown skull was the oldest human fossil yet found on British soil.

Woodward formally announced the find of the skull fragments and the jaw at a meeting of the Geological Society of London on December 18, 1912. He claimed the fossils represented a previously unknown hominid species, named it *Eoanthropus dawsoni* (*Eoanthropus* meaning "dawn man"), and proposed the new species might be the missing evolutionary link between apes and early humans. The basis for Woodward's view of the fossils was simple: The skull fragments appeared to be from a human-like skull (although thicker than present human skulls), while the jaw fragment (assumed to be from the same individual but with no evidence of articulation) was apparently ape-like. In short, the Piltdown individual was the earliest species of human lineage known at that time. One week later, on December 28, 1912, *The Illustrated London News* featured a full-page artist's reconstruction of the complete head of the Piltdown man, the sudden best candidate for the "missing link" between apes and humans.

But the reception at the 1912 Geological Society meeting was evidently mixed. Did the ape-like jaw really belong to the cranium? The prevailing theory was that in the evolutionary transition from ape to human, the large canine teeth of the ape evolved into the smaller canines of the human jaw. If the jaw was at all a human jaw, the canines should be smaller. Unhappily, there was no canine tooth (only an empty canine tooth socket) in the jaw fragment at hand.

Woodward agreed that the validity of the reconstruction could be solved once and for all only by finding the missing lower canine tooth that once existed in the part of the jaw that had been discovered.

So Dawson, Woodward, and Teilhard de Chardin returned to the Piltdown site and began digging again, and on August 30, 1913, it was Teilhard de Chardin who discovered the missing lower canine tooth that was taken to confirm the cranial and jaw fragments as belonging to the same individual, a primitive human.

The Piltdown man immediately became an international celebrity. The Piltdown site was already a tourist attraction, with highly publicized visits by many luminaries, including Arthur Conan Doyle, the creator of Sherlock Holmes. The press had found a ready source of entertainment for the public.

The Piltdown discovery appeared to confirm that the larger human brain evolved before other human physical characteristics, and that something close to the modern human form appeared extremely early in the human lineage, an idea of great importance to many prominent British anthropologists, who had been proposing this for some time and considered it the British view of things.

One result of the Piltdown discovery was that Neanderthal man was disqualified as a direct ancestor to *Homo sapiens*, since Neanderthal fossils, although more primitive in form than Piltdown, appeared in more recent geological beds than Piltdown.

Both the British public and British scholars were apparently also happy with the idea that the British Isles had evidently been an important site of early human evolution.

It took forty years to complete, but eventually the Piltdown discovery was transformed by evidence into the Piltdown hoax. First, in 1925, the geologist F. H. Edmonds declared the Piltdown gravels to be much less ancient than supposed because the height of the gravel bed above sea level was considerably less (102 versus 120 feet) than first reported by Dawson. Then, in the 1930s, discoveries in Europe, Asia, and Africa completely isolated Piltdown man as an evolutionary entity. Finally, after World War II, when new techniques were available, and attention returned to the Piltdown find, the Dawson-Woodward-Teilhard de Chardin "missing link" unravelled completely.

The human cranium was demonstrated to be modern, only approximately six hundred years old.

The jaw and teeth were demonstrated to be from an orangutan.

And the lower canine tooth found by Teilhard de Chardin (carefully filed to have a shape in accord with expectations) was apparently from a chimpanzee.

All the Piltdown fossils had evidently been fabricated and deliberately planted in the shallow gravel pits at Piltdown.

Microscopic study of the molars of the lower jaw revealed scratch marks made by filing of the teeth to simulate the human flattening of teeth produced by wear from chewing.

The skull and jaw appeared to have each been deliberately broken into the pieces discovered, the jaw fragment missing the part that would articulate with the skull to demonstrate connection.

Who did it has remained a mystery. Was it Dawson alone? Or Dawson and Woodward together? Or those two joined by Teilhard de Chardin? A recent (1996) suspect has been Martin A. C. Hinton, who became keeper of zoology at the British Museum in 1936 and who in 1912 had been working as a volunteer at the museum. The idea is that Woodward had rebuffed Hinton's request for a weekly wage, and Hinton, who already had considerable knowledge and skill with fossils, planted the Piltdown fossils to embarrass Woodward. In contrast, the paleontologist and evolutionary biologist Stephen Jay Gould (1941–2002) has made a strong case in print that the hoax was a conspiracy between Dawson and Teilhard de Chardin, Gould's arguments convincing enough to make this the most reasonable hypothesis.

But who perpetrated the hoax is of less importance than the major consequence of the hoax: The Piltdown fraud delayed for many years recognition of the significance of hominid fossils discovered in Africa, and completely distorted ideas about the significance of the Neanderthals. The hoax affected the work of an entire generation of anthropologists.

In a commentary on the Piltdown hoax, the anthropologists Richard E. Leakey and Roger Lewin wrote:

> The Piltdown forgery illustrates the sometimes indecent eagerness with which scientists will accept what they want to believe. Researchers today are not exempt from this weakness, and it can be seen in all branches of science. But because theories in [anthropology] are often constructed from relatively little data, in that field the danger of over-interpretation and therefore biased theories is particularly acute.

British anthropologists at the turn of the twentieth century made a serious error by overvaluing British contributions to their field. Anton Chekhov (1860–1904), already dead before the Piltdown hoax, put it well: "There is no national science just as there is no national multiplication table; what is national is no longer science."

ABDERHALDEN'S FERMENTS

THE CATALYTIC ACTION OF CERTAIN biological substances was discovered in the early nineteenth century, long before any understanding of their basic chemistry. What was known at that time was that these entities, called "ferments," could in small amounts accelerate the breakdown of natural substances such as starch. Later, at the end of the nineteenth century, it was found that some ferments could act as catalysts for the synthesis of natural products.

The systematic nomenclature of enzymes was developed in the latter half of the nineteenth century without knowledge of their molecular identities. Willy Kuhne (1837–1900) suggested in 1876 that the "ferments" found inside yeast cells be called "enzymes," from the Greek "in yeast." This name gradually replaced the older term "ferments," although the older term continued to be used by many researchers well into the twentieth century. E. P. Duclaux (1840–1904) suggested in 1883 that the suffix "-ase" be used in the names of enzymes to indicate their character, deriving this from the name "diastase," the name given to the first enzyme isolated in 1833 by Anselm Payen (1795–1871) and Jean-François Persoz (1805–1868).

In general, the chemical nature of enzymes remained unclear until 1930, when the American chemist John Northrop (1891–1987) crystallized pepsin and it soon became evident that all enzymes studied at that time were proteins. (Fifty-two years passed before the discovery of natural non-protein enzymes. The catalytic action of ribonucleic acid as a "ribozyme" was first discovered by Thomas Cech in 1982.)

One of the most interesting cases of scientific fraud in the twentieth century, and a case not widely discussed in the press or in scientific journals, concerns the enzyme research of the clinical biochemist Emil Abderhalden (1877–1950). Abderhalden was a scientist of outstanding reputation, professor of physiology and physiological chemistry at the University of Halle from 1911 to 1950, president of the oldest German academy of science, the Leopoldina, from 1911 to his death in 1950, the editor of several journals (including a journal on ethics), and author of several books and more than a thousand research papers. From any viewpoint, Abderhalden was an eminent scientist, discoverer of what he considered his major life work, what he called "defense enzymes," human catalytic proteins that were the subject of hundreds of his own papers and thousands of papers by others in the early part of the twentieth century. The problem is that defense enzymes do not exist, did not ever exist, and the consensus among the world's clinical biochemists today is that all of Abderhalden's work on these enzymes was fabricated junk science.

Abderhalden, born in Switzerland, was a physician who became a clinical biochemist. He received his medical training at the University of Basel, and in 1902 he started work in the laboratory of the Nobel laureate organic chemist Emil Fischer (1852–1919). In Fischer's laboratory, Abderhalden worked on the synthesis of peptides, the monomers of proteins, and on the action of proteases, the protein enzymes that break down other proteins. In 1909, Abderhalden published his first work on what he then called "protection enzymes" (Schutzferments), later called "defense enzymes" (Abwehrfermente). In 1912 he published a book on this subject, and three new editions of the book appeared before 1914. Clinicians in Europe and the United States scrambled to apply the new "discovery."

According to Abderhalden, the defense enzymes are produced when the body is invaded by foreign protein. Of great clinical significance were his reports that the serum of pregnant women contained proteases specific to proteins of the placenta. The idea was that the pregnant mother formed in her blood defense enzymes against fetal proteins. The test was simple: Placenta was boiled, and the denatured protein treated with serum from a pregnant woman. Peptides produced by the action of the specific defense enzymes were then dialyzed and identified by the biuret or ninhydrin reactions (two tests for proteins or peptides).

This pregnancy test was considered of great importance by gynecologists and biochemists, and by 1914 nearly five hundred papers on defense enzymes had appeared in scientific and medical journals, most of them German publications. The test was claimed by Abderhalden and his followers to be useful in other contexts besides pregnancies: the diagnosis of sarcomas and other cancers, the diagnosis of infectious diseases such as syphilis, and the diagnosis of psychiatric diseases such as schizophrenia.

In 1913, the biochemist Leonor Michaelis (1875–1949), who had just published his classic paper with his assistant Maud Menten on enzyme kinetics, was asked by the director of the hospital where Michaelis worked to verify the

Abderhalden test. Michaelis spent a week in Abderhalden's laboratory at the University of Halle, but he was unable to confirm Abderhalden's pregnancy test: Michaelis found none of the claimed differences between the sera of pregnant and nonpregnant women, or between women and men. Abderhalden's pregnancy test did not work. If anyone else had failed to confirm the test, they evidently hadn't publicized it. When Michaelis and his collaborator published their negative results in 1914, it apparently marked the end of his academic career in Germany: His prospects for a senior position vanished. He subsequently emigrated to the United States, where he worked for many years at the Rockefeller Institute.

The American biochemists Donald van Slyke (1883–1971) and Florence Hulton also were unable to verify Abderhalden's results. By 1920, the Abderhalden reaction was apparently no longer in use or even mentioned in the United States. In the journals he edited, however, Abderhalden, now president of the Leopoldina, the highly regarded German academy of science, continued his myth of defense enzymes in serum.

This is an ugly story with political tangents. Abderhalden was a eugenicist, and between 1933 and 1945, he held his high posts with the approval of the Nazi regime. In 1942, the geneticist Otmar von Verschuer was appointed director of the Kaiser Wilhelm Institute for Anthropology in Berlin, and von Verschuer's former assistant Josef Mengele joined him at the institute in the winter of 1942–1943. When, in April 1943, Mengele moved to the Auschwitz concentration camp to become camp doctor, von Verschuer applied for a grant from the Deutsche Forschungsgemeinschaft to finance Mengele's work in Auschwitz on defense enzymes. The project was to study defense enzymes produced by members of various races who were to be deliberately infected with various infectious diseases. The idea was to use the Abderhalden reaction to demonstrate racial differences, and von Verschuer accordingly sent a technician to Abderhalden's laboratory to learn Abderhalden's technique from Abderhalden himself. Von Verschuer appointed the biochemist Gunter Hillman to supervise the Abderhalden tests that would be made on blood sent from Auschwitz. Mengele subsequently sent blood samples from Jewish and Gypsy twins to the von Verschuer research group, and the group began analyzing the samples. In October 1944, von Verschuer reported in correspondence: "Precipitates have been prepared from the plasma of more than 200 individuals of various races, some twin pairs, some families."

After more than forty years of German work on defense enzymes, thousands of published papers on defense enzymes, and concentration camp inmates deliberately infected with fatal diseases to produce defense enzymes, the current consensus among biochemists is that defense enzymes never existed and that the results obtained by Abderhalden and his collaborators were at best wishful thinking and at worst completely fabricated.

SO NOT ONLY WERE the defense-enzyme medical experiments in concentration camps horrific in themselves as violations of human rights, but they were based on junk science and revealed no information of use to clinical biochemistry.

How was this enormous extended fabrication possible? To understand this case, and similar cases in certain sciences, one must consider the following general rule about fabrication of data in science: It's usually difficult for a young scientist, or a relatively unknown mature scientist, to publish a great deal of fabricated data about an important subject, since the experiments will soon be attempted by others and the fabrications revealed. In contrast, it's usually easy for a researcher of high status, an authority figure, to fabricate and publish over an extended period of time, since the majority of researchers will automatically assume the work is legitimate. Presented with experimental results obtained by a scientist of some renown, researchers in the same field almost always assume the validity of the data and move on to the new questions posed by the results. The system works, for the most part, and because of it science is able to sometimes move forward in large steps, with researchers leap-frogging each other in tandem as new ground is broken. But when the system doesn't work, when extensive fabrication by a respected authority occurs, the consequences can be catastrophic.

Abderhalden succeeded in duping researchers in his field because he was an authority figure, a member of the elite of biomedical science in a society where authority figures were considered to be beyond criticism.

Writing about the Abderhalden case in 1998, the German scientists U. Deichmann and B. Muller-Hill state:

> The elite of today [the biomedical elite in Germany] are loyal students of the old elite, and they have learned and internalized the old values. Has medical clinical science in Germany today really changed that much? We doubt it. The Brach-Herrmann-Mertelsmann affair [see below] provides a brief glimpse into the abyss of medical science in Germany. Will it be soon forgotten by the German medical elite, or will there be real change in the spirit of true science?

A SCANDAL IN MOLECULAR MEDICINE

THE BRACH-HERRMANN-MERTELSMANN AFFAIR MENTIONED ABOVE by U. Deichmann and B. Muller-Hill is a scientific fraud scandal in biomedicine that rocked the German science community in the past decade—nearly half a century after the death of Abderhalden, but with similarities in the way scientific elitism prevented a quick exposure of scientific fraud. This new scandal in Germany received much attention from the international science press, and its final outcome disturbed many researchers both in Germany and elsewhere.

The German scientific community was first stunned by the Brach-Herrmann-Mertelsmann scandal in the spring of 1997. The case involved two German molecular biologists, Marion Brach (who soon resigned as a professor at the University of Luebeck) and Friedhelm Herrmann (soon a suspended professor at the University of Ulm). Brach admitted falsifying published data while she worked under the supervision of Herrmann at the Max Delbrueck Center for Molecular Medicine in the 1990s. Herrmann claimed he was only a clinician and had nothing to do with the laboratory bench work.

The two researchers were accused by a whistle-blower, Eberhard Hildt, who worked in their research group, Hildt stating Brach and Herrmann had forged several research papers in hematology during their time at the universities in Mainz, Freiburg, and Berlin. The research involved the use in gene therapy of cytokines, low-molecular-weight proteins with hormone-like action that are secreted by various cell types and act as chemical messengers, particularly in the immune system.

Herrmann was a prominent hematologist and a leading genetic therapy researcher, and at the time the scandal broke, his salary was apparently US$278,000 per year. The two biologists had worked together for many years, and each received professorships on the basis of that work. When the scandal surfaced, all of that work was investigated, and the German government promptly established a commission of international scientific experts to discuss research standards and the procedures for scientific oversight in Germany.

Within a few months, Herrmann announced that he was suing various academic investigators in the case for DM10 million (US$5.6 million) in compensation for damage to his career. And at the same time, a new investigating team claimed that the two accused researchers also published falsified data earlier in their careers.

In the spring of 1998, Brach published a letter in the journal *Nature* in which she pointed out the following :

1. She had confessed to falsifying scientific papers.

2. She had resigned her position as a full professor at the University of Luebeck.

3. She did not believe further victimization was appropriate.

4. She had concluded that the various German investigating commissions met only with the intention of limiting the damage to the German academic community rather than with the intention of discovering the full extent of culpability.

5. The German government had reneged on its legal agreement to provide her with severance pay following her early confession and resignation.

6. Official bodies had found it expedient to imply that she was the major or only culprit in the affair.

During 1998, further details became public. Marion Brach and Friedhelm Herrmann had worked and lived together during their scientific collaboration at Harvard, Freiburg, and Berlin. They were apparently two lovers engaged in an alleged systematic fabrication of data in thirty-seven publications over a nine-year period ending in 1996. The press headlined the scandal as the worst case of scientific fraud in Germany since 1945. Many people in the international science community were astonished. These were two apparently accomplished scientists in a hot field, gene therapy, each with an important professorship at a major German university. If scientists of this caliber fabricated data, who could be trusted?

Brach evidently left Germany in 1998 and she was reported to be working in New York. Herrmann resigned his professorship at the University of Ulm and began working in private medical practice in Munich, while continuing to deny any involvement in misconduct and stating that the failure to bring charges against him was proof of his innocence.

An independent task force had been set up by the DFG (Deutsche

Forschungsgemeinschaft), Germany's main research granting body, and in the year 2000 the task force reported that 94 scientific papers (of Herrmann's 347 published papers) contained falsified data or data manipulation, the papers published between 1988 and 1992. Only 132 publications were cleared of any suspicion of fraud.

The scandal had now spread to the head of the department in which the Herrmann laboratory was located. Roland Mertelsmann, a well-known cancer researcher, and the head of the Department of Oncology and Hematology at the University of Freiburg, was publicly criticized by an investigative panel for failing to detect the alleged data falsification and data manipulation that occurred in his department. Mertelsmann, in fact, was listed as coauthor in fifty-eight of the Herrmann papers.

The investigative panel, headed by Albin Eser, director of the Max Planck Institute for Foreign and International Criminal Law, also cited Mertelsmann for "serious irregularities" in two Mertelsmann papers not related to Herrmann: a September 1994 paper in the journal *Blood*, and an August 1995 paper in *The New England Journal of Medicine*. The panel had evidently found that some data in these papers, which report on clinical trials of cancer treatments, were presented to give an impression of being "more complete and consistent than was actually the case." Mertelsmann has called the inquiry unfair and indicated he would mount a vigorous defense.

In the year 2002, the members of the task force that had been set up to investigate the affair publicly accused the DFG of watering down their report. "By playing down striking manipulations of data, the DFG has departed altogether from its role as keeper of good scientific practice."

In the early part of 2004, the case was turned on its head when the German government announced that Herrmann was to be allowed to keep his professorship at the University of Ulm. The district attorney of Berlin and Herrmann's attorney had reached an agreement that the case would not be taken to court after all, and that Herrmann could no longer be called a "forger" in the German press.

The district attorney of Berlin and Herrmann's attorney agreed on a minor fine of US$10,000, the government to cease further legal investigation into the case. According to current German law, Herrmann's guilt was labeled as "negligible," with repetition of the offenses unlikely, and the public interest not significant, since Herrmann had apparently not obtained his research grants on the basis of the fabricated data.

What was this case about? The case involved no erroneous assumptions or twisted conclusions—it was simply a matter of fabricated data, fabrications to make possible reports of "new" research results, which in turn brought lucrative prestige to the researchers. Prestige is the commodity of exchange in science, and researchers need publications of original work in order to acquire prestige and claim its rewards of increased salaries, lucrative professorships, and more funding for research aimed at acquiring even more prestige. If the researchers are honest, the self-interest at least produces legitimate and often important new science; if the researchers are dishonest, the self-interest gives the public only junk science, useless and often dangerous.

Many people have complained about the way the German government handled this case. Was there a whitewash? Standing apart from the intricacies of the politics of science and law in contemporary Germany, what is striking about this case is that nearly one hundred scientific publications by a single research group, publications in leading international journals, were found by the task force to contain fabricated data, the instigation of the case not a failure to replicate experiments by other laboratories, but instead the accusation of a whistleblower within the research group itself. If Eberhard Hildt had not blown the whistle on his research group, would the fabrications still be unrevealed? The case is a salient example of how authoritative researchers can publish fabricated data in many scientific papers over an extended period—all without discovery.

THE CASE OF THE FAKE FISH

THERE IS SOMETIMES AN ELEMENT of farce in revealed scientific fraud, since in hindsight it may seem ludicrous that the perpetrator or perpetrators of the fraud expected to get away with it. But farce aside, there are serious aspects to scientific fraud, first the damage to the ideals of the scientific community, and second (and maybe more important) the damage caused to other scientists by the fraud provoking costly strategic decisions based on the faked evidence.

The case of the fake fish concerns a living fossil—a teleost fossil. In general, the term "teleost" refers to any of the bony fish, the most advanced in terms of evolution and the largest group of fish currently extant. Besides the calcified internal skeleton, the most obvious uniform characteristic of the teleost fish is its tail, with upper and lower halves of about equal size, whereas in cartilaginous fish the tail has two lobes of unequal size. Almost all sport, commercial, and ornamental fish are teleosts.

We need some markers here in the history of life on Earth. The Devonian period is a geologic time frame that extends from approximately 400 million to 345 million years ago. It's sometimes called the "Age of the Sea," since more of the Earth was underwater than now. The Cretaceous period is the geological period ranging approximately from 146 million years ago to 65 million years ago.

The coelacanth is a primitive teleost fish that apparently first appeared in the Devonian period, with the most recent fossil specimens dating from the Cretaceous period. The group was assumed to be extinct, but a living specimen was caught in 1938 in the mouth of the Chalumna River in South Africa and named *Latimeria chalumnae*. The 1938 specimen was two meters in length and weighed forty kilograms. More specimens have since been caught, all off the coast of South Africa and near the island of Madagascar (Comoros archipelago) in the Indian Ocean.

In July 1998, M. V. Erdmann and colleagues reported the capture and observation of a live coelacanth specimen near the island of Manado Tua (north Sulawesi) in Indonesia, and the find was quickly published in the journal *Nature* and recognized as of considerable importance. In the published article, a photograph of the captured coelacanth appeared, a lateral view of the complete fish.

In the spring of the year 2000, three French scientists (Bernard Seret, Laurent Pouyaud, and Georges Serre), one of whom (Seret) was an ichthyologist at the Museum of Natural History of Paris, submitted to the journal *Nature* an article claiming a prior discovery in 1995 of a coelacanth in Indonesian waters (thus preempting the 1998 discovery of M. V. Erdmann et al.), the article submitted by the Seret group including a photograph of their captured fish. The French group stated they were unable to register their specimen in 1995 because it failed to reach the museum to which it had been sent. They stated they photographed the fish at the time, then lost the photograph while moving house, and only found the photograph again in the year 2000.

In addition to the question of priority of discovery, the scientific implications of the findings of the French team were that it would extend the distribution of living coelacanths, since the French fish was stated to have been caught more than 2000 kilometers from the spot where Erdmann et al. found the 1998 specimen, which suggests a larger distribution of the fish in the Indo–West Pacific region.

When staff people at the journal *Nature* noticed that the photograph submitted by the French group led by B. Seret et al. and the photograph by M. V. Erdmann et al. previously published two years before were virtually identical, the Seret et al. paper was refused by *Nature*, and the French photograph was denounced as an outright fake. Apparently, computer graphics software was used to simply cut and paste the 1998 Erdmann et al. fish into another photograph, moving the coelacanth onto a table with two other ordinary fish as though all three fish were in the same catch. When presented with the hard evidence that the two coelacanth photos were identical, the Seret et al. response was that the picture was "taken by a friend who later died and whose widow gave it to Serre before moving abroad." In the July 13, 2000, *Nature* issue, the apparently faked photo and the original photo published in 1998 by Erdmann et al. appeared side by side for comparison.

Again, aside from the farcical aspects here (clipping a photograph of an unusual object from a journal article and using the clipped photograph to fake another photograph submitted as "evidence" in another article sent to the very same journal), it's obvious that a bit more cleverness with computer software might have made the fake unrecognizable as such, and steered coelacanth specialists onto a false trail for decades. That is not amusing.

Commenting on this affair in a journal letter, M. V. Erdmann and R. I. Caldwell (University of California, Berkeley, U.S.), state: "The Indonesians and Comorans are rightfully proud of efforts in their two countries to preserve these rare and very special fish. What pride can we in the Western scientific community take in this affair?"

AN AFFAIR IN PHYSICS

So FAR, THE MODERN CASES we have discussed have been in the biological sciences. In the fall of 2002, the physical sciences community was startled by its own scandal. A series of extraordinary and highly publicized advances at the venerable Bell Laboratories was declared by an outside investigating committee to be based on fraudulent data.

The committee's findings essentially dismissed as fiction results from more than a dozen published research papers that had been touted as major breakthroughs in physics, including claims that Bell Labs had created molecular-scale transistors. The research had been widely publicized in general science magazines and in consumer magazines devoted to science.

The researcher involved, Jan Hendrik Schoen, a German native and considered a "star" investigator in microelectronics, was fired after the outside investigating committee discovered he had been falsifying experimental data.

Schoen received his Ph.D. from the University of Konstanz in Germany in 1997. In December 2000, he was hired by Bell Labs.

In the period from 1998 to the summer of 2001, Schoen was listed as an author on an average of one research paper every eight days. In 2001 he announced in the journal *Nature* that he had produced a molecular-scale transistor. He claimed to have used a thin layer of organic dye molecules to assemble an electric circuit that, when acted on by an electric current, behaved as a transistor. Research into the electrical properties of organic films was already highly developed in many laboratories. Schoen's results implied a significant step forward to a shift away from silicon-based electronics and toward organic electronics, with the possible shrinking of transistors into a size domain prohibited by standard silicon chips. Sharp reductions in the costs of chip manufacture were also envisioned as a consequence.

But soon after Schoen began publishing his results, other researchers in the physics community noted in public that Schoen's data contained anomalies: The data seemed too precise, and some of the results apparently contradicted fundamental physics.

In this context, "noise" is electrical noise, and electrical noise is thoroughly random and temperature dependent. Because of its random nature, fundamental physics says the noise recorded in any two experiments can never be identical, and that's especially true if the two experiments involve different temperatures.

Lydia Sohn at the University of California, Berkeley, noticed that two of the Schoen experiments, carried out at very different temperatures, had identical noise—a contradiction of basic physics. When the editors of the journal *Nature* pointed this out to Schoen, he said he had accidentally submitted the same graph twice. Paul McEwan of Cornell University then found the identical noise in another Schoen paper describing a third experiment. Further research by Sohn, McEwan, and other physicists revealed a number of examples of duplicate data in Schoen's reports. In total, twenty-five papers by Schoen and his twenty coauthors were considered suspect.

In May 2002, Bell Labs appointed Malcolm Beasley of Stanford University to chair a committee to investigate possible scientific fraud in Schoen's published work. The committee sent questionnaires to all of Schoen's coauthors and interviewed his three principal coauthors. The committee examined electronic drafts of the suspect papers, drafts that included processed numeric data. They requested copies of raw data, but found that Schoen had kept no laboratory notebooks. His raw data files had been erased from his computer. According to Schoen, the files were erased because his computer had limited memory. In addition, all of his experimental samples had been discarded or damaged beyond repair.

The investigating committee concluded that Schoen, thirty-two years old, had concocted or altered data at least sixteen times between 1998 and 2001 — the first known case of scientific fraud in the seventy-seven-year history of Bell Laboratories. The laboratory, once a part of AT&T, is currently the research arm of Lucent Technologies. The investigating committee found that whole data sets were reused by Schoen in a number of different experiments. They also found that some of his purportedly experimental results had been produced using mathematical functions. The fabricated findings had been published in several prominent journals, including *Science, Nature,* and *Applied Physics Letters.*

The investigating committee cleared of charges approximately twenty other researchers from Bell Labs and other institutions who worked on the research or helped write reports about the results. The investigating committee stated, "The evidence that manipulation and misrepresentation of data occurred is compelling," and that Schoen "did this intentionally or recklessly and without the knowledge of any of his coauthors."

On October 31, 2002, the journal *Science* withdrew eight papers written by Schoen. On March 5, 2003, the journal *Nature* withdrew seven papers written by Schoen.

In a response appended to the report, Schoen stated that he disagreed with several of its findings, but "I have to admit that I made various mistakes in my scientific work, which I deeply regret." Schoen blamed some mistakes on the work's complexity or errors he did not notice before publication. But he said all the scientific publications he prepared were based on experimental observations.

Researchers at Delft University of Technology and the Thomas J. Watson Research Center subsequently performed experiments similar to Schoen's, but could not obtain similar results.

In June 2004, the University of Konstanz deprived Schoen of his doctoral degree.

So what was this case about? Schoen losing his doctoral degree was only a personal consequence, but not the essence of the case. As far as understanding junk science is concerned, the important aspect is that both Bell Laboratories and the international physics community were fooled until someone noticed that noise records published by Schoen in several papers were identical — which means physically impossible. If Schoen had been a shade more clever or energetic, he could easily have doctored the noise to make the records differ. Ultimately, his experiments would have been replicated by others and questions might have been raised. In how many years? A few years? A decade? Detecting fabricated data in the physical sciences may be generally easier than in the biological sciences, but how often does detection fail?

MYTHS ABOUT FRAUD IN SCIENCE

THERE ARE TWO PREVALENT MYTHS concerning scientific fraud.

The first myth states that since most scientific experiments are replicated by other laboratories, science is self-correcting because the discovery of fraud involving the fabrication of data is inevitable.

The second myth is that scientific papers involving fabrication of data are extremely rare, with only a few fraudulent papers published in any one year.

Concerning the first myth, it's not true that all or even most published experiments are sooner or later replicated. What is true is that research results that are of apparent great significance will probably be replicated quickly, but not other research results, and especially not other results produced by the laboratories of authorities in the field. These days, in the front lines of fast-moving fields in science, few laboratory heads are inclined to waste precious manpower and funds simply replicating the work of others. The most successful research strategy is to assume the relevant published product of other laboratories is honest and take the next step, or better yet, the step after that, in the push to solve a hot problem.

Concerning the second myth, the evidence that exists suggests the number of scientific papers involving fabricated data published each year may involve hundreds and perhaps thousands of publications.

For example, the Office of Research Integrity (ORI) of the U.S. National Institutes of Health (NIH) publishes yearly reports of discovered fabrication of data by NIH grantees. Between 1993 and 1997, sixty-one researchers receiving its grants were involved in the fabrication of data. The total number of NIH grants during the same period was approximately 150,000. These figures produce a fabrication rate for published papers of 0.0004, assuming each research grant produced on average the same number of papers. This fabrication rate involves the biomedical sciences in the United States (the area with which NIH is concerned), but I'm not aware of any evidence of a lower rate in other areas of science.

NIH grantees usually publish in leading journals. In 1998, the top 5,600 scientific journals published approximately 1 million articles, and it's estimated that this is approximately one-tenth of the total world output. Considering only the top 1 million publications and applying the fabrication rate derived from the NIH evidence, we find the expected number of publications each year in all scientific fields involving fabricated data is approximately 400 papers for the top 5,600 journals, and perhaps as much as ten times that if all scientific journals are considered. It's certainly true that these estimates involve some assumptions; but it's also true that a small fabrication rate applied to a large number of published articles suggests a significant number of fraudulent publications.

Another approach to the question of the magnitude of the problem is to consider what working scientists apparently know: In one survey, published in *American Scientist* in 1993, between 6 and 9 percent of scientist respondents said they were personally aware of results that had been plagiarized or fabricated within their faculties. Take the middle percentage (7.5 percent) of science faculties in the world, and the apparent result is the existence of enough corrupted installations to produce thousands of fraudulent research papers each year.

In general, the idea that only a few fraudulent scientific papers are published each year is probably a complete fallacy. What is true is that few frauds are exposed and therefore known to the science community—with even fewer known to the public.

Does it matter? You might think the problem is esoteric, a problem for scientists rather than for the public. But these days the gears of our society depend on science: Our medicine, our engineering, our national security are all based on the work of scientists. In the twenty-first century, science is more a public endeavor than ever before, and corrupted science is a public problem.

PART TWO

BUYER BEWARE

3

THE FOOD AND DIET CIRCUS: EAT YOUR WAY TO HAPPINESS

Kissing don't last: cookery do!
— GEORGE MEREDITH (1828–1909)

You have a situation where girls of eight want to lose weight and at twelve they can tell you the fat content of an avocado. But they don't know what constitutes a healthy meal.
— MARY EVANS YOUNG

FOOD IS A FUNDAMENTAL human necessity, and nearly every aspect of food production and food consumption has been commercialized and regulated in one way or another. Commercialization is necessary because most of us don't grow our own food, and regulation is necessary because without it there's a danger that the food we buy will be contaminated or chemically altered. As with anything that affects nearly all of society, people debate constantly about methods and goals. Meanwhile, a number of questions are always in front of us: Do we eat too much? Are the dangers of cholesterol in foods real? Are certain foods unhealthy? What role do genes play in individual nutritional needs? Are dietary drugs safe? Are vitamin supplements and other food supplements necessary? Are the antibiotics used in meat production causing antibiotic-resistant strains of bacteria to proliferate? All of these questions are important, and maybe a dozen more could be added to the list.

These questions also all involve aspects of food production and food consumption that are vulnerable to junk science delivered to the public by industry and government to support special agendas. In this chapter I'll consider some important myths and realities about food consumption and food production and the junk science that has people confused about what and how to feed themselves and their children.

The classical circus has three rings; the food and diet circus has enough rings to make you dizzy.

THE FOOD MANIA

ARE WE REALLY OBSESSED WITH food? In a culture of abundance, eating food is transformed from a necessity into a sport. We're a people who have eating contests. We seat a dozen contestants at a table and the winner is the one who eats the most hamburgers in twenty minutes. We giggle as we watch the contestants hurry to stuff their faces with hamburgers.

We're told that food is a delight, abundant food our reward for living in a rich country. We're confronted by images of food in magazines, television, film, newspapers, food books, food shows, food channels. And pervading all of it is junk science so artfully developed, it's almost invisible. The game in the food marketplace is one of twisted science, exaggeration, withheld data, ballyhoo, and carefully designed hucksterism to exploit the ignorance of the public. And the public suffers for it.

History has its lessons about food and exploitation. In nineteenth-century urban America, when government regulations for the safety of food and consumables were almost nonexistent, rotting meat was sold openly to the poor. It was common knowledge among New Yorkers that their milk was diluted with water. Butter was often sold rancid to the poor, and often mixed with casein, water, calcium, gypsum, gelatin fat, or even mashed potatoes.

One of the definitions of the word "circus" is "a place of activity marked by rowdy or noisy behavior." The commerce and distribution of urban food in the nineteenth century was indeed a circus whose drumbeat was scarcity of food for the poor of the cities. There is more food available now, food aplenty and much of it cheap, maybe too cheap, since these days the prevailing "food problem" is a problem of overeating rather than hunger. At least this is true in the industrialized countries of the West; elsewhere in the world, hunger is still ordinary.

The international contrasts of scarcity and plenty are often stark and unsettling: A newspaper photograph shows a child in some undeveloped country dying of starvation, and on another page in the same newspaper one finds one more news report about the epidemic of overeating and obesity in America.

There is no dispute concerning the tragedy in the first case; but in the second case, the American public continues to vacillate between confused resolve and eager self-indulgence. A reader's letter to a major newspaper says: "At this point in our history we are able to enjoy eating. . . . It is difficult not to laugh at our own preoccupation with weight."

But Americans do want to look thin, and telling people how to diet is a lucrative business that keeps people preoccupied with various diets and schemes in books, TV, and magazines. Before-and-after pitches tell people they can drop as much weight as they like without heartache, effectively saying, "Just buy my book and I'll make you look wonderful in a bikini."

Were that the sum of it, a joy in eating contradicting a preoccupation with weight for the sake of the appeal of slenderness, the entire problem of the epidemic of obesity in the industrialized world would be no more than an ephemeral triviality. After all, the approved amount of body fat on a body, from the standpoint of fashion, changes markedly with time and culture. And we certainly enjoy eating, have apparently evolved to enjoy eating, no matter the body-shape fashion that may be in force at the moment. It's an old question, so old that it carries with it an ancient cautionary proverb: "Eat to live and not live to eat." Somewhere, someplace, many centuries ago, the conflict between the joy of eating and weight gain was already apparent.

Is it all so harmless?

Not so. In the twenty-first century, a new aspect has become part of the question, an aspect whose import has not yet taken root in the public mind: Overeating is not merely a matter of gluttony overriding current fashion, it's also a medical problem with tragic consequences, and a problem with a huge financial burden to the public in medical costs.

The problem with overeating is that it leads to obesity, and the problem with obesity is that it's the major risk factor for type 2 diabetes, a killing disease. So this is not a question of fashion or appearance or gluttony or restraint, it's a question of slow suicide aided and abetted by the media and food industry.

CONSIDER AN EXAMPLE OF industry distortion: The chairman of the National Automatic Merchandising Association (vending machines) writes a letter to the *Chicago Tribune* (February 12, 2005) and reports a study of purchases from vending machines at schools for more than a million schoolchildren. The chairman says, "The survey evaluated 2,084 school vending machines and concluded that students actually purchased far less than one candy bar or salted snack item per student per week . . . clear evidence that taking vending machines out of our schools will have little if any impact on what is in fact a very complicated problem."

But what is not mentioned is that if only a minority of students bought the bulk of candy bars and snacks, the per capita purchases within that minority could be ten or twenty times greater than for the population of children as a whole, and it's those children who become obese, with their obesity aided and abetted by school vending machines. If this is the case, is it a public service for the newspaper to print such a corporate distortion of reason without commentary under the headline HELPING FIGHT OBESITY IN CHILDREN?

The medical aspects of food and obesity are clear. Although a hundred years ago type 2 diabetes mellitus was considered a rare disease, the incidence of the disease has recently exploded in industrialized countries. There are currently approximately 16 million Americans with type 2 diabetes, and at least an equal number have impaired glucose tolerance, a precursor of diabetes. Insulin resistance (diminished effectiveness of insulin in lowering plasma glucose levels) and hyperinsulinemia (high levels of insulin in blood plasma) are characteristic of both type 2 diabetes and impaired glucose tolerance (excessive levels of blood glucose). These metabolic derangements, combined with the high blood pressure and abnormal blood lipids that are common in type 2 diabetes and impaired glucose tolerance, markedly increase the risk of cardiovascular, peripheral vascular, and cerebrovascular disease.

And this of course provokes the public health question: Why has the incidence of type 2 diabetes increased so rapidly? A large body of epidemiological evidence points to excess caloric intake and physical inactivity as the major reasons for the epidemic. A chronic imbalance between energy expenditure and energy intake causes excessive weight gain and obesity, and obesity is one of the most important risk factors for insulin resistance and type 2 diabetes.

Excessive caloric intake increases the mass of adipose tissue in the body, and adipose tissue is composed of adipocytes—fat cells. This cell type is no longer regarded as merely a passive depot for storing excess energy in the form of triglycerides. It's now apparent that adipocytes actively regulate the pathways responsible for energy balance, and that the activity of these cells is controlled by a complex network of hormones and neuronal signals. Adipocytes secrete chemical messengers that include leptin, tumor necrosis factor alpha, angiotensinogen, and adiponectin. The most recently discovered adipocyte-secreted hormone is resistin, which may be an important link between increased fat mass and insulin resistance.

Are the numbers as clear as the physiology? Between 1960 and 2000 the prevalence of obesity among adults aged 20 years to 74 years in the United States increased from 13 percent to 31 percent. An estimated 325,000 deaths and between 4 percent and 6 percent of direct health-care costs (approximately $39 billion to $52 billion) are attributed to obesity annually.

THE INCONSISTENCIES AND CONFUSIONS about food that are delivered to the American public by government bureaucrats and the media have made food consumption an unregulated minefield for the consumer.

If a certain food kills people within a few days after ingestion, the law forces the withdrawal of that food from the food market. If a certain food kills people (and is known to kill people) after ten or twenty years of repeated ingestion, the law does nothing to protect the victimized public.

On January 12, 2005, the U.S. government released its *Dietary Guidelines for Americans*, the report emphasizing the need to reduce total calorie intake. But where are the educational campaigns? Michael Jacobson, the head of the Center for Science in the Public Interest, a nutrition advocacy group, has warned that the fight against obesity requires educational campaigns and new legislation to subsidize healthy foods and curb food advertising aimed at children.

The huckstering of science is often subtle. As an example, consider a recent full-page advertisement for pistachio nuts in a "health" magazine. The ad shows a lovely young woman hovering over pistachio nuts, and the advertisement says:

> *Scientific evidence suggests but does not prove that eating 1.5 ounces per day of most nuts, including pistachios, as part of a diet low in saturated fat and cholesterol, may reduce the risk of heart disease.*

The amount of pistachio nuts indicated, 1.5 ounces, will provide 141 calories and 13 grams of fat, which are moderate quantities. The important part of the text is the phrase "as part of a diet low in saturated fat and cholesterol." Eating almost *anything* in moderate quantities as part of a diet low in saturated fat and cholesterol may reduce the risk of heart attack. Most nuts are healthy foods and a good source of protein, but they are generally high in calories per unit weight and there is no evidence that alone they reduce the risk of heart disease. It's the low–saturated fat, low-cholesterol diet that reduces the risk, not the nuts. The advertisement is a subtle distortion, the point of it to make people believe that eating pistachio nuts is a preventive against heart disease. The headline says, "When choosing a snack, go with your heart," with the word "heart" in large type to associate it with pistachio nuts, artful hucksterism twisting science in a so-called health magazine.

Not every food is healthy, and eating in excess is as dangerous as smoking cigarettes. Where are the surgeon general's warnings about food? Is public health at the mercy of the food industry?

The average energy in kilocalories expended per hour by adults depends on the weight of the person and the activity at the moment. A sleeping person weighing 140 pounds will burn 58 calories per hour or 464 calories in eight hours of sleep. With moderate activity, the number of calories burned per hour rises to 166, or 2,656 calories in sixteen hours of wakefulness, which means approximately 3,120 calories burned in a twenty-four-hour cycle.

These numbers aside, the important principle is that if more calories are taken in as fuel than are burned, the excess is stored by the body, with 3,500 calories of stored energy deposited as an additional 1 pound of weight. The result is that a small excess intake of only 100 calories a day will add approximately 1 pound a month to body weight, 12 pounds a year, and 60 pounds in five years. With only moderate but consistent overeating, the clear consequence is obesity.

Excess weight in children, for anyone who knows the dangers, the degree to which the child's future may already be compromised, is a sorrow to behold. The prevalence of excess weight doubled among children six to eleven years of age, and tripled among those twelve to seventeen years of age, between 1976 and the most recent survey in 2000. The current view in medicine is that excess weight in children can no longer be considered a benign condition or a condition related only to appearance. Between 1979 and 1999, the rates of obesity and obesity-associated diagnoses, such as sleep apnea and gallbladder disease, tripled among children six to seventeen years of age. Approximately 60 percent of overweight children and adolescents, in addition to being overweight, have at least one additional risk factor for cardiovascular disease, a risk factor such as elevated blood pressure, hyperlipidemia, or hyperinsulinemia.

❀   ❀   ❀

IN AMERICA, THE FLIP side of the coin of the food mania is the weight-loss mania. Public attention to the dangers of obesity predictably creates a natural market for commercial weight reduction schemes—weight-loss regimens and diets that if successful in the marketplace can bring great riches to their originators. As one segment of commerce pushes the public to overeating and obesity, another segment of commerce provides an array of weight-loss pills, regimens, and devices.

At present, the most popular diet schemes are the various low-carbohydrate diets touted by accomplished hucksters. There is insufficient evidence, however, to make recommendations for or against the use of low-carbohydrate diets for weight loss, particularly among people older than fifty years, for use longer than ninety days, or for diets of 20 grams per day or less of carbohydrates. Among the published studies, weight loss while using low-carbohydrate diets was for the most part associated with decreased caloric intake and increased diet duration, but not with reduced carbohydrate content. The general rule remains: To lose weight, eat less; there are no self-indulgent diet fixes that allow both weight loss and high caloric intake.

IS DIETARY CHOLESTEROL DANGEROUS?

THE COMMON BIOLOGICAL MOLECULE CHOLESTEROL surged into public awareness in the 1970s. First purified in 1934, cholesterol is the principal sterol of higher animals, found in all body tissues, especially in the brain, spinal cord, and in animal fats and oils. It is also the main constituent of gallstones.

Cholesterol is an essential molecule in all animals, an important constituent of all cell membranes in animals. There is no cholesterol at all in any plant or plant derivative. Apart from its role in animal cell membranes, the major clinical importance of cholesterol is that it's not soluble by itself in water (or blood plasma), it complexes with certain soluble proteins (so-called lipoproteins), and these complexes may come out of solution on the interior walls of arteries and ultimately produce inflammation and arterial blockage.

The body constantly synthesizes cholesterol, no matter how much cholesterol is contained in one's diet: The important consideration is how much cholesterol appears in blood, and that quantity is heavily dependent on both dietary cholesterol and on lipid metabolism. The concentration of blood cholesterol in any individual is a result of both individual biochemistry and dietary cholesterol input, and an individual's lipid-metabolism dysfunction, hereditary or acquired, can result in high blood cholesterol even with a cholesterol-free diet. Similarly, the degree to which high blood cholesterol produces pathological arterial changes is unique to each individual and apparently involves inflammatory mechanisms that are quite complicated.

Dietary cholesterol is indeed an important factor related to high blood cholesterol, but not as important as many food advertisers claim. Despite the junk-science ballyhoo about low-cholesterol foods, most dietary manipulations result in only modest cholesterol reductions of 4 percent to 13 percent, and diet has been considered by some researchers as a relatively ineffective therapy for high blood cholesterol levels. The problem is that in most cases individual biochemistry determines the effects of dietary cholesterol: Low-cholesterol diets do not produce the same results in all people. A low-cholesterol diet is not necessary for people with normal cholesterol levels, and it's not a sufficient remedy for people with high cholesterol levels.

In contrast, 3-hydroxy-3-methylglutaryl coenzyme A reductase inhibitors (statins such as Lipitor and Zocor) repeatedly have been shown to reduce mean serum low-density lipoprotein cholesterol (LDL-C) concentrations by 28 percent to 35 percent in long-term trials, with corresponding reductions in cardiovascular death of 23 percent to 32 percent in both primary and secondary prevention trials. But here also individual differences are important, since with certain individuals even moderate doses of statins may produce unwanted effects on liver function and muscle action.

NUTRITIONAL GENOMICS

COMPOUNDING THE PROBLEM OF FOOD and diet ballyhoo are the junk-science offers to the public of "nutritional genomics," the offers purporting, for a price, to provide an analysis of individual nutritional needs based on an individual's genome. Several companies offer genetically based nutrition advice. Clients who submit a cheek swab, plus responses to a lifestyle questionnaire, and $395 to a company called Genelex in Redmond, Washington, a distributor for the UK-based company Sciona, receive a customized report recommending nutritional changes that can "improve health and prevent illness." The recommendation is claimed to be based on genetic profiling of nineteen genes of the client and analysis of the questionnaire. Since experimental evidence relating genes to nutrition in healthy individuals is almost nonexistent, what is the genomic basis in this scheme for making individual recommendations? Is this a legitimate service or a junk-science scam based on the public's lack of understanding of current genomic science?

An article in *Newsweek* adds to the circus antics by puffing nutritional genomics: "Within a decade . . . doctors will be able to take genetic profiles of their patients, identify specific diseases for which they are at risk, and create customized nutrition plans accordingly. Some people will be advised to eat broccoli. . . ." One turns the page and finds the following: "Green tea helps silence genes that fuel breast cancer in some women. . . . Broccoli boosts genes that protect against heart disease. . . . Soybeans affect 123 genes involved in prostate cancer. . . . Turmeric suppresses genes that ratchet up inflammation."

The simple fact that the relatively small number of studies in this field only suggest rather than demonstrate interactions is never emphasized. The field of nutritional genomics is so new that by next year every one of the relevant studies may be contradicted by new research.

Is there any hope for nutritional genomics in the near future? Consider this: The human genome has about 30,000 genes controlling about 100,000 different proteins. Single genes can be involved with more than one protein, and single proteins can be involved with more than one gene, all of it producing a chemical interaction system of enormous complexity, the details of which are still mostly unknown. In the past decade we've elucidated the possible nutritional relevance of about a dozen genes and corresponding proteins, only a paltry number of the genes and proteins potentially relevant for nutrition. At this rate, nutritional genomics may be useful in the twenty-second century, but probably not useful until we get there. The current nutritional genomics hype is more about a possible application of human genome analysis than about nutrition: It's a way for entrepreneurs to make a buck out of the present gene craze.

From a journalistic standpoint, of course, one never emphasizes doubt, only certainty of dramatic ideas. The public is entertained, and if in the process the public is misinformed, the next issue will carry a new cover story and public attention will be somewhere else. For the media, continuing the noisy circus is more important than public understanding.

And so it goes. The modern public is bombarded, blasted, and bamboozled by food and diet advice and products, largely because most of these commercial activities are unregulated and exceptionally lucrative. People are tempted by food shows on television, then admonished to eat less, then advised what to eat by nutrition "experts," and then urged by food supplement suppliers to add various vitamin, mineral, and other pills to the diet in place.

THE DANGERS OF DIETARY DRUGS

THE STORY OF THE DIETARY drug Fen-Phen illustrates how junk science permeates the drug industry in its promotion of "off-label" use of pharmaceuticals.

The term "off-label" refers to use of a drug for treatments not approved by the Food and Drug Administration (FDA). In the United States, physicians are essentially autonomous, and when a drug is approved by the FDA for use in the treatment, for example, of epilepsy, any physician can write a prescription for the same drug to be used in the treatment of any other disease or condition, no matter what the evidence (or lack of it) supporting that use.

Fen-Phen is the common name of a drug combination that became the most widely sold pharmaceutical for weight loss in the 1990s, the drug selling during a virtual Fen-Phen craze that did not end until 1997, after two of the drugs used in the combination were linked to deadly lung and heart disease and were abruptly pulled off the market after killing more than a hundred people. The manufacturer of Fen-Phen has apparently paid $13.6 billion in damages in a negotiated settlement.

The drugs used in Fen-Phen are fenfluramine (or related dexfenfluramine) and phentermine. Fenfluramine (Pondamin) is a halogenated amphetamine that produces a profound reduction in levels of 5-hydroxytryptamine in the brain and a consequent loss of appetite. The drug was first approved by the FDA as a central nervous system amphetamine pharmaceutical in 1973 for use in psychiatric treatment. The second drug used in the Fen-Phen cocktail, phentermine, was approved much earlier, also for psychiatric treatment, in 1959, and has a central nervous system action similar to amphetamines. Beginning in 1992, reports began appearing in medical journals suggesting that a combination of phentermine and fenfluramine resulted in significant weight loss in psychiatric patients when used over an extended period of time.

But Fen-Phen, the two drugs in combination, has an important side effect:

primary pulmonary hypertension. This affects the heart valves, causes them to leak, which in turn causes regurgitation of blood back through the valves, a consequence that can be fatal. The FDA never approved any of the studies suggesting the Fen-Phen drug combination would help weight loss. The drugs were marketed for off-label use by drug salesmen coaching physicians, and when heavy consumer advertising quickly caused an enormous consumer demand, physicians began writing prescriptions for a weight-loss drug that had not passed through adequate clinical trials to ensure its general safety as a weight-loss pharmaceutical.

No doubt many of the hundred or more people killed by the Fen-Phen cocktail as they attempted to lose some weight thought they were protected by the U.S. Food and Drug Administration and the expertise of the physician who wrote their prescriptions. Unfortunately, the FDA had not approved the cocktail for weight loss and the physicians were practicing junk medicine, medicine based on junk science. Any physician who writes a prescription for a drug when he or she has no evidence that the drug is not only effective but safe and has only a primitive understanding of the possible interactions of the drug is practicing junk medicine, discussed again in Chapter 7.

Ephedra (ephedrine) is another weight-loss drug, an amphetamine-like nervous system stimulant used as an herb for centuries in China. In the United States, ephedra is a common ingredient in numerous dietary supplements. This drug stimulates the central nervous system much like adrenaline and caffeine, and it can have potentially dangerous effects on the nervous system and the heart. The health hazards associated with ephedra include high blood pressure, irregular heartbeat and heart palpitations, nerve damage, seizures, heart attack, stroke, and death. Ephedra is listed on ingredients labels variously as phedra, ephedra sinica, ephedra extract, ephedra herb powder, Ma Huang or Ma Huang extract, epitonin, ephedrine, Chinese ephedra, pseudoephedrine, and norpseudoephedrine.

Because ephedra products are commonly marketed as dietary supplements, there is little or no pre-market review by the FDA of their safety (or effectiveness), and proper dosing information is not available to the consumer.

Not all popular weight-loss pills are related to prescription drugs. One of the most popular weight-loss supplements currently sold in the United States is Zantrex-3. The pill sells for approximately $50 for a one-month supply, contains a large dose of caffeine, a dash of green tea, and three common South American stimulant herbs. That's all. Nothing else is in the pill, and there's no evidence that the simple ingredients directly affect weight loss any more than a cup of coffee. Zantrex-3 first appeared on the American market in March 2003 and has been a huge commercial success with millions of bottles sold in vitamin shops and chain drugstores such as CVS, Rite Aid, and Wal-Mart. Zantrex-3 is only one of an enormous repertoire of diet supplements and weight-loss pills, thousands of different tablets, elixirs, potions, and pills sold in the United States—and very little is known about most of them.

The safest and surest way to lose weight is to eat less and exercise as much as possible: Cut down on fuel intake and burn fuel as much as you can. Taking some unregulated and untested pill or potion into your body makes you vulnerable both medically and financially.

VITAMINS AND OTHER FOOD SUPPLEMENTS

VITAMINS ARE ESSENTIAL ORGANIC COMPOUNDS that cannot be synthesized by humans and therefore must be ingested to prevent metabolic disorders. Although classic vitamin deficiency syndromes such as scurvy, beriberi, and pellagra are now uncommon in Western societies, specific clinical subgroups remain at risk. For example, elderly patients are particularly at risk for vitamins B_{12} and D deficiency; alcohol-dependent individuals are at risk for folate, B_6, B_{12}, and thiamin deficiency; and hospitalized patients are at risk for deficiencies of folate and other water-soluble vitamins. Inadequate intake or subtle deficiencies in several vitamins are risk factors for chronic diseases such as cardiovascular disease, cancer, and osteoporosis. But aside from the evidence concerning specific vitamin deficiencies and specific pathologies, there is much ambiguity in the scientific literature concerning the benefits of vitamins as food supplements to an already adequate diet. At least 30 percent of U.S. residents use vitamin supplements regularly, which makes the production and sale of vitamin supplements a billion-dollar industry and one with little or no regulation to protect the public.

Since 1994, when Congress deregulated the food supplement industry and opened the market to a flood of new products, the marketplace has seen an enormous increase in the use of dietary supplements, weight-loss supplements, and various potions that are touted to do everything from provide high energy to improve memory to prevent old age. Since the Dietary Supplement Health and Education Act became law in 1994, companies can say nearly anything they please about the potential health benefits of what they sell—without providing any scientific evidence—as long as they do not explicitly lie or claim a cure for a specific disease. There are almost no standards that regulate how food supplement pills are manufactured or what they contain, and no proof is required that the products are effective or safe before they are put on the market. Is this an example of how the U.S. Congress protects the health of the American consumer? Why not go all the way and return to the free-market unregulated jungle of the nineteenth century, the jungle that sold to consumers cocaine in Coca-Cola, arsenic in milk, and maggots in meat? The food and diet circus thrives on public ignorance of science and medicine, and since this ignorance is not about to vanish by any magic wand waved anywhere, leaving this particular circus unregulated is an extreme disservice by Congress—a gross disservice to the American public.

THE DANGERS OF ANTIBIOTICS IN FOOD PRODUCTION

A PARTICULARLY INSIDIOUS FORM OF junk science, a form that will also be apparent in several later instances, occurs when industry or government, rather than overtly twisting science to suit an agenda, simply ignores the relevant science with an argument that existing evidence of public danger is not strong enough to warrant a change in policy. During the past thirty years, this has been the position of the U.S. government and many American meat and poultry producers concerning the use of antibiotics in food production, a position that implicitly twists science by denying the strength of evidence produced by sound research.

The problem of microbial multi-drug resistance to antibiotics has been developing for many decades, a result of the use of antibiotics in the meat and poultry industries to enhance the growth of farm animals used in food production. Bacteria in these animals, both pathogens and nonpathogens, quickly evolve strains resistant to these antibiotics and other antibiotics that are chemically similar, and this antibiotic resistance is passed not only to the progeny of these bacteria but to other species of bacteria as well, the passage an example of gene transfer.

Biologists recognize two types of gene transfer from one organism to another: vertical and horizontal. Vertical gene transfer occurs between parents and offspring, and horizontal gene transfer is the transfer that may occur between organisms otherwise.

It's in bacteria that horizontal gene transfer has been studied most extensively, particularly in the last decade, and it's horizontal gene transfer that is apparently involved in the dissemination of antibiotic resistance throughout the general bacterial population.

The production of antibiotic resistance in bacteria is a food industry phenomenon. Approximately 8 billion animals are raised each year in the United States for human food consumption, most of these animals receiving some antibiotics during their lifetime. These include 7.5 billion chickens, approximately 100 million turkeys, 1 million cattle, and 1 million pigs. This animal population provides an enormous reservoir of antibiotic-resistant bacteria contaminating soil, farm implements, and farm workers. Evidence indicates that whole farming communities are contaminated with antibiotic-resistant strains of bacteria, and of course this contamination is passed into the general human population in many ways.

The problem is not the pathogens inside meat and poultry that comes to the kitchen: With proper cooking, those pathogens are eliminated (although if the food is handled before cooking, contamination in the kitchen is certainly possible). The major problem is the antibiotic-resistant pathogens that are passed into the general population from the contaminated meat and poultry industries, from their implements, from the farm soil contaminated with animal feces, from industry workers, and from any community in contact with these farming operations. The end result is an infiltration of antibiotic-resistant strains of bacteria into the entire human population, an infiltration that has evidently produced disastrous consequences.

An illustrative case is that of the salmonella pathogen. Salmonella (named after the pathologist Daniel E. Salmon [1850–1914]) is a genus of gram-negative rod-shaped bacteria, various species of which cause food poisoning (salmonellosis), typhoid fever, paratyphoid fever, and some forms of gastroenteritis in humans.

Salmonella is a leading cause of food-borne illness, with 1.4 million cases of salmonellosis occurring each year in the United States. Most salmonella infections in humans result from the ingestion of contaminated poultry, beef, pork, eggs, and milk. The consensus among microbiologists is that the emergence of antimicrobial-resistant salmonella is associated with the use of antibiotics in animals raised for food, and that resistant bacteria can be transmitted to humans through foods, particularly foods of animal origin.

In 2001, D. G. White and colleagues identified and characterized strains of salmonella isolated from ground meats purchased in the Washington D.C. area. Salmonella were isolated from samples of ground chicken, beef, turkey, and pork purchased at three supermarkets, and the isolates were characterized by various techniques. Of two hundred meat samples, 20 percent contained salmonella. Of these, 84 percent were resistant to at least one antibiotic, and 53 percent were resistant to at least three antibiotics. Sixteen percent of the isolates were resistant to ceftriaxone, the drug of choice for treating salmonellosis in children. The conclusion was that resistant strains of salmonella are common in retail ground meats. The researchers suggested that national surveillance for antimicrobial-resistant salmonella should be extended to include retail meats.

Antibiotic-resistant bacteria of other types are also apparently widely disseminated throughout the United States, and the situation is severely compromising medical use of antibiotics to the point where it is expected that for certain bacterial diseases there will soon be no antibiotic treatment available. Examples are staphylococcus infections, enterococcus infections, shigella dysentery, and typhoid fever. The reality is that any bacterial species has the potential to quickly evolve antibiotic-resistant strains when consistently exposed to antibiotics.

What can be done? Until 1970, the use of antibiotics to promote growth in food animals occurred in the UK and other parts of Europe. The British government appointed a commission of microbiologists and physicians to investigate the practice, and in the issued UK Swann Report this commission concluded that the practice was harmful to human health, that antibiotics administered to animals for prolonged periods produced a strong selection for resistant bacteria in the animal intestinal flora, and that these bacteria are a potential risk to the human population. The commission already had evidence that the pathogen salmonella could be traced from animals to disease in humans. Use of antibiotics to enhance animal growth was subsequently banned in England, other European countries, and Canada. No such legislative ban has ever been issued in the United States. How many people in the United States must become ill or die before something is done?

FOOD IS AS NECESSARY a part of our life as the air we breathe. But food and everything related to it are also huge businesses that have always thrived on ballyhoo and junk science. Can people be confident that their government is protecting them against the hucksters and barkers in the food and diet circus? If not, why not? The next time you hear about lobbyists working Congress for special interests, don't push it aside as an abstraction. The "working" of Congress by lobbyists is what keeps the food and diet circus in town—and we all pay for it with our health and money.

4

GM FOODS: FRANKENSTEIN IN
A CORN PATCH

GM FOODS ARE PLAYING GAMES WITH NATURE
— HEADLINE IN THE *LONDON DAILY MAIL,* FEBRUARY 1999

Science, with technology, is the only way we have to avoid
starvation, disease, and death.
— MAURICE WILKINS

IN THE SUMMER OF 1999, the public in Europe and the United States
found itself inflamed over the issue of genetic modification of food crops
and food animals. In the UK, militant groups engaged in acts of violent van-
dalism and succeeded in destroying valuable farm acreage and expensive research buildings devoted to genetic engineering of plants. In the United States,
the protests were not as intense as those in Europe, maybe due to the preoccupation of the American public with the personal life of President Clinton, but
there were American protests aplenty, and many of them were disruptive.

What started this frenzy? Although it's a question for sociologists, it seemed
apparent at the time that the media enjoyed fueling an issue whose basic idea—
altered food—most people could understand, and an issue that would bring
readers to the magazines, newspapers, and television talking-head illuminators.

People readily eat fake butter containing chemicals they have never heard
of, but try selling them broccoli with a single gene changed to protect the plant
against insects. Truly, there is something holy about genes, something sanctified even in a plant. Either that or the public enjoys having nightmares about
killer tomatoes. I think it's safe to choose the nightmares as a causal factor: The

public enjoys nightmares, and the media make money providing the public with nightmares at every opportunity.

Of course, not all nightmares are paranoid fantasies. The trick is to keep your feet on the ground and demand attention to reality. Are genetically modified (GM) foods a monster created by biotechnology? Are Frankenstein and his sidekick Igor wandering in our cornfields? What is the reality?

BIOTECH ON A RAMPAGE

THE MAGAZINE *TECHNOLOGY REVIEW* IS published by the Massachusetts Institute of Technology, an institution whose closest rival for academic technological celebrity is the California Institute of Technology on the other coast. Since biotechnology, if anything, is technology supreme, an impacting new technology, one would expect the official magazine of the Massachusetts Institute of Technology to at least have an attitude that reflected an understanding of the difficulties of any new technology that impacts society. Unfortunately, no matter the expectations, reality has its own devices.

The cover of the July/August 1999 issue of *Technology Review* blared BIOTECH GOES WILD, the headline on a background of metallic-looking plants shaped into barbed wire. Inside the magazine (which was then in its one hundredth year of publication), the lead article of the same title was by a noted science writer, Charles C. Mann. What was striking was that the text itself was even-handed, while the title and blurbing appeared deliberately alarmist: "Biotech Goes Wild" was also the title of the article.

The lead blurb: "Genetic engineering will be essential to feed the world's billions. But could it unleash a race of 'superweeds'? No one seems to know. And nobody's in charge of finding out."

Another blurb: "The worry is that biotech crops will spontaneously breed with wild relatives, creating 'superweeds.'"

The editorial in the issue, by editor-in-chief John Benditt, was as even-handed as the article, which made both the magazine cover and the blurbs in the Mann article even more surprising. Maybe the explanation is that MIT's *Technology Review* is a newsstand publication, and the cover was deliberately designed to be provocative. Defenders of the magazine might point out that since the actual content of the article was even-handed, the content would counteract the effect of the cover. Unfortunately, most people who approached a newsstand that summer and saw the cover of *Technology Review* did not buy the magazine or even open it: They were merely left with BIOTECH GOES WILD ringing in their heads, reinforcing other misinformation and fears of the unknown.

It's ironic that in the context of the science of genetics, the term "wild" denotes the "normal" and unmodified plant. In the UK that summer, the tabloids called genetically engineered agricultural products "Frankenstein Foods." Years before that summer, MIT's *Technology Review* had been a publication seriously devoted to accurate communication of new science and technology to the public. Apparently, that focus shifted in the summer of 1999.

Several months later, a front page article appeared in *The New York Times* (November 3, 1999). The lengthy article, in addition to the front page paragraphs, occupied a full internal page and was concerned with genetically engineered agriculture and foods, in particular with a squash plant genetically engineered to be resistant to destructive plant viruses. The headline of the article announced SQUASH WITH ALTERED GENES RAISES FEARS OF "SUPERWEEDS."

Of course, the provocation of the headline, which was all most people would read, is a common modus operandi in newspaper journalism: In the absence of a cause for fear, there is little reason to write about the squash plant or food production, certainly no reason that would sell newspapers. *The New York Times* article, as did most popularized accounts of the subject of genetic modification of foods, gathered quotations from various sources to support the notion that extreme concern about genetic engineering in agriculture was warranted—in other words, to sustain and provoke public fear. Little space in the article was devoted to counterarguments, or to the actual and potential benefits of genetic engineering in agriculture. The gist of the article was an emphasis on risks, with words like "may," "could," "might" appearing in nearly every paragraph.

Since *The New York Times* is the most influential newspaper in the United States, and since certain other U.S. media had been publishing similar articles, the prospect in 1999 was that the United States would soon have the problem already experienced in the UK—a strong, vocal, and militant protest movement against genetically engineered agriculture and foods—against so-called Frankenstein foods.

The criticism did not stop with newspapers. Also in the fall of 1999, there appeared a new book by bio-Luddite Jeremy Rifkin warning the world of an impending doom to be caused by genetic engineering. According to Rifkin, molecular biologists and the biotechnology industry were dismissing scientific studies that indicated the biotech revolution would likely be accompanied by the proliferation and spread of "genetic pollution" and the "wholesale loss of genetic diversity." Rifkin warned of the uncontrollable spread of superweeds, the buildup of resistant strains of bacteria and new "super insects," the creation of novel viruses, the destabilization of whole ecosystems, the "genetic contamination" of food, and the steady depletion of the "gene pool." Rifkin proclaimed:

> To ignore the warnings is to place the biosphere and civilization in harm's way in the coming years. Pestilence, famine, and the spread of new kinds of diseases throughout the world might yet turn out to be the final act in the script being prepared for the biotech century.

In the face of such junk science, the year 1999 was not a good year for the public image of biotechnology.

The sociological energy loop is apparent: The media provokes public fear, which in turn produces public interest in the subject, which in turn produces more media provocation, and so on, with an ongoing result an increase in media market share. In the instance of genetically engineered agriculture, the emphasis is always on risk: "may," "could," "might"—provocation by articulation of negative possibilities.

One must agree that there is risk. There is always risk in any new technology, precisely because the technology is new. In the past, this was true for the first explorations of the human species with fire, with the wheel, with the use of metals, with sea travel, with the introduction of the steam engine, the automobile, immunization, air travel, television, computers, the Internet, and so on. We are what are because in the past we took risks.

In 1829, the governor of New York, Martin Van Buren (1782–1862) wrote the following to President Andrew Jackson (1767–1845):

> As you well know, Mr. President, "railroad" carriages are pulled at the enormous speed of 15 miles per hour, by "engines" which in addition to endangering life and limb of passengers, roar and snort their way through the countryside, setting fire to the crops, searing the livestock and frightening women and children. The Almighty never intended that people should travel at such breakneck speed.

Never mind New York Governor Van Buren's ridiculous belief that he knew the intentions of the Almighty, let us focus on the risk assessment that must have occurred in the corridors of power. Yes, this was new technology; yes, the engines frightened both people and animals; yes, there could never be any certainty about all the negative consequences. But the potentially positive consequences were enormous, the tying together of the entire continent, the expansion of trade and industry, the reduction of the hazards of migration to the frontier—all of these had to be in the minds of the people in government who pushed for development of a railroad system on the American continent. Certainly, our present civilization and our economic strength depend on the risk-takers of the past.

But the most important point about risk, a point almost always avoided by the media, is that it's not possible in principle to prove a null hypothesis—in particular, it's not possible to prove that any drug or vaccine or food or technology is harmless, i.e., risk-free. All that one can do is demonstrate that under certain experimental circumstances there is no apparent harm; one cannot make a proof of no harm for all circumstances.

For example, it has not been proved (and it cannot be proved) that the inks now used by *The New York Times* and other print media are not carcinogenic. All inks contain dye molecules, many dyes are known to be carcinogenic, and many substances interact with biological cells at extremely small concentrations. Should we provoke public fear concerning these media inks? Is it in the public interest to do so? The article in *The New York Times* about genetically modified foods essentially castigated the U.S. Department of Agriculture for not requiring agricultural companies that produce genetically modified products to prove their products are harmless. Since no such proof is possible, the article in *The New York Times* was what is called a "bad rap" against the U.S. Department of Agriculture—a bad rap based on a junk-science belief that it's possible to prove a null hypothesis.

The year 1999 was a chaotic year for genetic engineering: strife, chaos, and a variety of bad raps against science, agricultural business interests, government agencies, and even against the militant groups themselves, who were sometimes wrongly accused of being anarchists.

The chaos in Europe was greater than the chaos in the United States. A survey in 1999 concluded that no single explanation could account for the greater resistance in Europe than in the United States to food biotechnology. Various factors seemed implicated and interrelated. Different histories of media coverage and regulation coalesced with different patterns of public perceptions, and these in turn reflected deeper cultural sensitivities, not only toward food and novel food technologies but also toward agriculture and the environment.

But the fundamental theme of the chaos was clear: Many people did not like the idea of biotechnology, and specifically genetic engineering, applied to the foods they ate, and the concern soon spread to foods that other people ate, that the whole world ate—forecasts of a doomsday produced by mad scientists running amok in the corn fields and wheat fields of the world. The chaos was all about food.

BIOTECHNOLOGY AND NUTRIENT PRODUCTION

THE CONSUMPTION OF FOOD (NUTRIENTS) is one of the main activities of all living systems, and the development and improvement of production of nutrient sources has been one of the main activities of the human species throughout its history. Agricultural researchers generally believe that the application of new techniques in molecular biology and genetics to agriculture will be of great importance to our species in the twenty-first century, and indeed that application is already underway and essentially constitutes the beginnings of a revolution in the technology of nutrient production.

These days, one focus of opposition to change is the introduction of new methods in agricultural technology, particularly the application of genetic engineering techniques to crop production.

At the present time in human history, the prime movers in applied science are in the commercial sector, profit-making enterprises that generally put up the capital required for applied science in order to bring a return on the investment to their shareholders. That's our current system, and in the context of agricultural technology, it's to be expected that those corporate enterprises responsible for the application of genetic engineering to agriculture will bear the brunt of vocal opposition generated by fears of technological innovation. One corporate enterprise that has received a good deal of that brunt is the Monsanto Company, a leader in the application of genetic engineering in agriculture worldwide. In the last decade, in the United Kingdom, protesters wrecked experimental crops designed to improve food production. More than 175 years before, in 1815, in the same country, protesters—called "Luddites," after leader

Ned Ludd—wrecked the new machines of the Industrial Revolution designed to improve the production of cloth and other goods.

New technologies rarely slip peacefully into the workings of civilization.

GENE TECHNOLOGY

WHAT ARE THE SCIENTIFIC REALITIES concerning genetically modified foods? The simplest reality is that modification of the genes of plants to improve crops is a practice that has been in place since the dawn of human history. What is new is merely a faster way to do it. The new gene technology speeds up the process, allows more control of details, makes possible totally new agricultural systems—but the end result, gene modification by genetic "engineering," is the same end result as that achieved by thousands of years of crop manipulations (selective breeding) by farmers.

Throughout the centuries, farmers have used selective breeding to alter the genetic profile of food plants. With careful selection and breeding of phenotypes, one can push the genetics of a population of plants in one direction or another. Nature (evolution) does this extremely slowly, usually in time frames of millions of years. Traditional farming does it much faster, usually in centuries or decades.

But one limitation of selective breeding is that it relies on the farmer's ability to carry out genetic crosses between individual plants, and for this such plants must be sexually compatible, which in turn means they need to be closely related. The result is that with selective breeding it's not possible to combine genetic traits from widely different species of plants, or to introduce new traits not found in plants at all.

On the other hand, genetic engineering allows the introduction of any trait known to be associated with one or more genes, the introduction accomplished by essentially splicing new genes into the old genome.

The current major method of introducing new genes into the genome of plants involves bacteria. *Agrobacterium tumefaciens* is a pathogenic bacterium that causes cancerous crown gall disease in plants, a disease common in a variety of plant species. The bacterium provokes tumors in plants by transferring part of a large bacterial DNA segment to its plant hosts. Once within the plant cell, this transferred DNA moves to the plant cell nucleus, where it eventually integrates into the host genome. Expression of the new genes (in the natural case, the cancer-producing oncogenes) results in the loss of cell growth control and consequent tumor formation. Genetic engineering uses this naturally occurring transfer mechanism by substituting one or more genes of interest for the oncogenes in the bacterium, and thereby achieving entry of these new genes into plant cells. Once transferred, these "transgenes" can endow plants with new characteristics, including herbicide resistance or pathogen resistance, altered growth or nutritional qualities, or the ability to produce drugs, edible vaccines, essential nutrients, and so on. This "horizontal gene transfer" between living kingdoms, between bacteria and plants, was once thought to be an exclusive property of the bacterium *Agrobacterium tumefaciens*, but other bacterial species have recently been shown to be capable of the mechanism.

So there is the gist of it: A common bacterium that causes tumors in plants by injecting its own tumor-causing genes into the plant genome is genetically engineered by humans to instead inject human-selected genes into plants, human-selected genes that produce plants more useful to the human species. Though the method is new, is there anything really "bizarre" about genetically modified food crops?

Science keeps moving us away from the apes, our knowledge of the world widening the distance between us and other primates. Of course, if one wants to be an ape, one objects to the movement, maybe with charming speeches about the benefits of "natural man." Try making such a speech to a ten-year-old child going blind from a lack of carotene, the precursor to vitamin A, when, as we will see later on, we can put that carotene into the rice that child eats daily.

GM CROPS AND ECOSYSTEMS

THE TERM "ECOSYSTEM" REFERS IN general to a community of organisms interacting with each other, plus the environment with which they live and with which they also interact. The study of ecosystems is the study of interacting components, animate and inanimate, in a particular region of the Earth's surface—or on a global scale.

The arguments voiced against genetically modified crops come from many quarters and represent various fears, and ecosystem destruction is often suggested as a potential disastrous consequence of man-made interference. But in truth there is no possibility for human life on Earth without "man-made" interference with a variety of ecosystems.

As ecologists continually point out, all organisms modify their environment, and humans are no exception. A bird catching a worm is modifying its environment, just as a human mowing a lawn or tossing a beer can into a ditch is modifying his or her environment. As the human population has grown and the power of technology has expanded, the scope and nature of human modifications have changed drastically. Until recently, the term "human-dominated ecosystems" produced images of agricultural fields, pastures, or urban landscapes; now the phrase can be said to describe the entire planet. Many ecosystems are dominated directly by humanity, and no ecosystem on Earth's surface is free of pervasive human influence.

The growth of the human population, and growth in the resource base used by humanity, is maintained by various human enterprises such as agriculture, industry, fishing, and international commerce. These enterprises transform the land surface through cropping, forestry, and urbanization; alter the major biogeochemical cycles; and add or remove species and genetically distinct populations in most of Earth's ecosystems. Many of these changes are substantial, and all are ongoing. These changes in turn cause further changes in the functioning of the global Earth system, especially by driving global climatic change and causing irreversible losses of biological diversity. The human species changes the Earth, has been changing the Earth since the dawn of human history, and since our existence depends on it, we can be thankful that the changing of the Earth will not stop.

THE GREEN REVOLUTION

THE SO-CALLED GREEN REVOLUTION THAT lasted from 1950 through the mid-1980s was one of the great technological success stories of the second half of the twentieth century. Due to the introduction of scientifically bred higher-yielding varieties of rice, wheat, and maize beginning in the 1960s, overall food production in the developing countries kept pace with population growth, with both more than doubling. The benefits of the Green Revolution reached many of the world's poorest people. Forty years ago there were a billion people in developing countries who did not get enough to eat, equivalent to 50 percent of the population of these countries. If this proportion had remained unchanged, the hungry would now number over 2 billion—more than double the current estimate of around 800 million, or approximately 20 percent of the present population of the developing world. In addition, world food prices have declined in real terms by over 70 percent since the 1970s. Those who benefit most are the poor, who spend the highest proportion of their family income on food.

Critics of the Green Revolution maintain that although agricultural technology did increase the total amount of yield per acre, the increased productivity caused degradation of the land, chemical contamination of ground water, sterilization of the soil, salinization and water-logging by irrigation, and erosion of top soil by tillage practices. This local environmental damage may have indeed resulted in places, but how many millions of people are alive today because the Green Revolution allowed them to avoid starvation as children, and how many of these children will mature to solve the local environmental problems caused by the new technology?

Some critics also maintain that world hunger is not a problem of food supply but a problem of poverty. The answer to this is that of course people who are not poor can afford to buy food. But increasing food production reduces the price of food, and with advanced technology we'll be able to increase food production long before we can eradicate poverty. In short, let's prevent starvation first, and take on the challenge of eradicating poverty later on.

GM FOODS AND BLINDNESS IN CHILDREN

ANOTHER CONSEQUENCE OF THE FEAR of genetically modified foods is even more dramatic: Holding back the use of genetically modified foods also holds back prevention of blindness in millions of children throughout the world. Each year 350,000 children die and another 2 million go blind because of vitamin A deficiency.

There are two forms of vitamin A deficiency. The primary form is caused by insufficient beta-carotene (provitamin A) in the diet. The compound beta-carotene is a precursor to vitamin A and is converted to vitamin A in a healthy body. The secondary form of vitamin A deficiency occurs as the result of a variety of metabolic dysfunctions that prevent the conversion of beta-carotene to vitamin A, including dysfunctions produced by general malnutrition.

Vitamin A deficiency causes symptoms ranging from night blindness to the symptoms of xerophthalmia and keratomalacia, leading to total blindness. In Southeast Asia, it's estimated that a quarter of a million children go blind each year because of this nutritional deficiency. Furthermore, vitamin A deficiency exacerbates afflictions such as diarrhea, respiratory diseases, and childhood diseases such as measles. It's estimated that improved nutrition could prevent 1 million to 2 million deaths annually among children. Oral delivery of vitamin A is difficult to manage, mainly due to the lack of societal infrastructure in undeveloped countries. We need alternatives to oral administration; it may be sufficient in industrialized countries, but it's not practical in undeveloped countries. One answer is the supplementation of a major staple food such as rice with provitamin A. Because no rice crop produces this provitamin, recombinant genetic engineering technologies ("gene splicing") are necessary—conventional plant breeding cannot do the job.

After much effort, researchers have genetically engineered a rice strain that produces beta-carotene, a precursor of vitamin A. This genetically modified crop, called "yellow rice" because of its color, could help the 124 million children worldwide who suffer from primary vitamin A deficiency.

In general, even where starvation is prevented, the world is still faced with the problem of seriously inadequate nutrition in large numbers of people. The social costs produced by consequent disease and early deaths are enormous and debilitate every undeveloped society. Researchers in the field therefore recognize that modifying the nutritional composition of plant foods is in general an urgent worldwide health issue, since basic nutritional needs for much of the world's population are still unmet. Large numbers of people in developing countries exist on simple diets composed primarily of a few staple foods (cassava, wheat, rice, and corn) that are poor sources of some macronutrients and many essential micronutrients. Consequently, the diet of over 800 million people does not contain sufficient macronutrients, and micronutrient deficiencies are even more prevalent. Estimates place 250 million children at risk for vitamin A deficiency (in up to 500,000 children a year, this deficiency will cause irreversible blindness); 2 billion people at risk for iron deficiency (with children and women of reproductive age particularly vulnerable); and 1.5 billion people at risk for iodine deficiency.

Biotechnology can improve trace element nutrition. For example, genetic engineering can be used to increase the trace element content of staple foods such as cereals and legumes. This may be achieved by introduction of genes that code for trace element-binding proteins, overexpression of storage proteins already present, or increased expression of proteins that are responsible for trace element uptake into plants.

The composition of oils, proteins, and carbohydrates in seeds of corn, soybean, and other crops has already been modified to produce grains with enhanced value. Both plant breeding and molecular technologies have been used to produce plants carrying the desired traits. Genomics-based strategies for gene discovery have accelerated the identification of product candidates for genetic modification. Molecular breeding strategies have been used to accelerate the process of moving trait genes into high-yielding germplasm for commercialization. These products are being tested for applications in food, feed, and industrial markets.

THE PROBLEM OF RISK ASSESSMENT

THE PRIMARY ISSUE THAT FEDERAL agencies and scientists alike must address in this context is that of risk assessment. As plant geneticist Paul F. Lurquin has pointed out, plants engineered for insect resistance may or may not harm other insects that are not pests for these plants. One must also be concerned about the potential spread of artificially introduced genes (transgenes) by sexual crossing between engineered crops and their related species. In addition, some people are concerned that GM plants might one day outcompete indigenous species, thereby deeply altering natural environments. Lurquin concludes that these questions are difficult to address, but they are not intractable. For example, a threat to innocuous insects can be evaluated in the laboratory under realistic conditions. Similarly, the spread of transgenes can be measured in greenhouse experiments and small-scale plots. Further, the fitness of GM plants and their ability to outcompete other plants also can be measured under controlled conditions.

In general, geneticists and agricultural scientists are well aware of potential negative consequences; it's a fallacy to believe these researchers are blind to the risks of genetic engineering of food crops.

And medical scientists are also concerned. There are concerns about the safety of genetically modified crops, concerns that they may contain allergenic substances due to introduction of new genes into crops. Another concern is that genetic engineering often involves the use of antibiotic-resistance genes as "selectable markers," and this could lead to production of antibiotic-resistant bacterial strains that are resistant to available antibiotics. This could create a serious public health problem. The genetically modified crops might contain other toxic substances (such as enhanced amounts of heavy metals) and the crops might not be "substantially equivalent" in genome, proteome, and metabolome compared with unmodified crops. Another concern is that genetically modified crops may be less nutritious; for example, they might contain lower amounts of phytoestrogens, which protect against heart disease and cancer.

All of these concerns are valid but involve only possibilities. People live with possibilities, including the possibility of getting slammed by a truck on your way to work, or the possibility of having your house slide down a cliff on the California coast. Risk assessment is the art of weighing possible gains against possible losses, an art already known to ancient humans before they ventured out of their caves every morning.

THE PRESENT SITUATION CONCERNING SAFETY

WHAT, THEN, ARE THE REALITIES of genetic engineering of the food supply? We are now seven years down time's road from the summer of 1999. A substantial fraction of the American food supply now involves genetic engineering in one form or another. Are we being choked by "superweeds"? Are new viral diseases, originating in genetically engineered plants, killing people everywhere? Has genetic engineering of food crops created Frankenstein insects to bedevil the public?

Is it possible that enough time hasn't elapsed and that any of these calamities might still occur? Again, nearly anything is possible. The important question is not what's possible, but what's likely?

The central fact is that the genetically modified crops now available in the marketplace that are intended for human consumption are generally safe; their consumption has never been associated with serious health problems.

And it's not just genetically modified foods that we've already accepted in the marketplace. As Richard Lewontin, no fan of agribusiness, pointed out in a recent review of the anti-genetic-engineering movement, the human gene for insulin has been successfully inserted into the genome of bacteria, and these bacteria, cultured in industrial vats, are now producing large amounts of human insulin for the market. Despite public fears about human ingestion of the products of genetic engineering, bio-Luddites seem to ignore the large number of diabetics who twice a day inject into themselves insulin produced by genetically engineered bacteria. Lewontin concludes: "As far as anyone knows, no one has been harmed by this product of genetic engineering; but then, as far as anyone knows, no one has yet been harmed by any product of genetic engineering."

Of course, a genetically modified food catastrophe "may" happen tomorrow; serious damage "might" happen the day after that; and next week genetically engineered corn plants "could" organize and decide to march on the nearest city. Meanwhile, this particular public controversy has been an exercise in media manipulation, antiestablishment frenzy, and junk science.

There are no intelligent reasons to stop or even slow down the application of modern biology to agriculture. Whatever cautions may be necessary for environmental protection are already in place among agricultural scientists, and if there are any unforseeable consequences, they can be dealt with when they occur. There is no scientific basis for predicting a biological doomsday caused by genetic engineering of food crops.

The real danger to the public is clear: Media-fed public opposition to genetically modified foods and opposition to research on genetically modified foods severely curtails improvements in food production to feed the hungry world and curtails the development of crops specifically designed to treat certain deficiency diseases.

We've had too many years of attacks against GM foods, alarms about killer weeds, giant insects, and rampant plagues. It's been a Luddite party based on ignorance, fear-mongering, and a good deal of careerism. Maybe it's time protesters against GM foods devoted their energies to cleaning junk science out of their barns.

5

AGING AND LONGEVITY: LIVE FOREVER NOW

Human growth hormone is not the fountain of youth, but it's the closest we have come to finding it.

—From HGH-PRO, a commercial
human growth hormone Web site

There is no current "magic bullet" medication that retards or reverses aging. . . . Going to the gym is beneficial and certainly cheaper than growth hormone.

—Mary Lee Vance, M.D., a leading
growth hormone researcher

AN ARGUMENT CAN BE made that the human attribute that most drives human affairs is awareness of individual mortality, an awareness that pervades all human interactions, strategies, personal choices, all of human history and politics, and all of the arts.

The reality that one's life has a limited, even ephemeral, duration is unpalatable at best and terrifying at worst, and any notion, scheme, or vague promise of immortality or even of significant life extension is grasped and held tight to the chest with both arms. We're like poor rats caught in a sealed room filled with slowly (or quickly) rising water, panic always in the wings of our minds, and so since the beginning of human history the process of aging and death has been a focus of myth, mysticism, and junk science.

People who grasp whatever is at hand in the hope that it will slow or stop the rising water are not to be faulted or derided. But there are those who exploit this vulnerability to achieve or maintain power, or for financial gain, who exploit with junk science and do great harm in the process.

THE SEARCH FOR LONGEVITY

ALTHOUGH HUMANS HAVE BEEN FASCINATED with the so-called fountain of youth for centuries, I will not deal here with the myths about aging and longevity before the era of modern science. We have enough such myths in our own time to keep us busy. Let's begin with the following statement in 1964 by Alex Comfort, a biologist trained in medicine: "If we kept throughout life the same resistance to stress, injury, and disease which we had at the age of ten, about one-half of us here today might expect to survive in seven hundred years' time."

A striking statement, but there was no evidence to warrant that statement in 1964 and there is no evidence for it today. Comfort's idea was based on assumptions concerning declining "power of self-adjustment and self-maintenance" in humans and other animals—processes which most biologists today would call homeostasis and tissue repair. These two processes are certainly involved in health and disease, and to the degree that health and disease determine life span these processes are also involved in longevity. But to what degree are they involved and how is a life span of seven hundred years arrived at? Comfort's conclusion was cavalier when he wrote it and remains so today. Unfortunately, his book was widely read around the world and used by others to tout an enormous variety of pills, potions, nostrums, and schemes to improve resistance to stress, injury, and disease and thus extend the human life span by centuries into the future.

HAS ANYTHING CHANGED IN the past forty years? What has changed is the vocabulary, the ballyhoo now with a new emphasis on computers and machines, genes, DNA, hormones, miracle drugs, cryogenics, and so on. Here is a modern version of futuristic longevity delirium:

> In the twenty-first century the convergence of artificial intelligence, nanotechnology, and genetic engineering will allow human beings to achieve things previously imagined only in science fiction. Life spans will extend well beyond a century. Our senses and cognition will be enhanced. We will gain control over our emotions and memory. We will merge with machines, and machines will become more like humans. These technologies will allow us to evolve into varieties of "posthumans" and usher us into a "transhuman" era and society.

In the above excerpt, the author, James Hughes, says, "Life spans will extend well beyond a century." Later, in the same book, he's more specific:

> Life expectancy has dramatically increased in the last century as the developed countries built public health systems. In the twenty-first century we will achieve radical life extension, with indefinite life spans. We need to ensure that everyone has access to these treatments.

Here the cat is out of the bag: Life span will be extended not a mere 700 years, but indefinitely—our twenty-first century will bring us all immortality, provided "everyone has access to these treatments." According to this fantastical scenario, anyone now forty or fifty years old had better break their neck on a run for the latest life-extension pills and schemes, anything to ensure they will live long enough to start to live forever.

The expressed confidence in near-term change is remarkable, given that the majority of the American public does not accept the theory of evolution, that the federal government bans federally funded stem cell research, and that much of the public, including prominent neoconservative members of the establishment, views advances in science and engineering with suspicion and sometimes hatred. Indeed, if biology and biotechnology have expanded in importance to the public, the expansion has brought "bio-Luddites" to the stage to warn us that too much biological knowledge and know-how will bring us doom, including a few on the sociopolitical neoconservative fringe who insist that death should be preferred over life-extension, since the prospect of death and the melancholy of the prospect have generated great art!

But most people disagree that an extension of human life would be an evil. What nearly everyone wants is longevity, a life as long as possible, and since the public is without much sophistication in the biology of longevity, the longevity industry becomes a purveyor of junk science, the pitch always made with the flavor of fashion in modern science and current medical biology.

HORMONES AND AGING

FOR MORE THAN A CENTURY, hormones have been public celebrities as chemicals that control growth, mood, sex, food intake, and a host of other basic physiological functions, apparently including some aspects of the aging process.

And of all the chemical substances found in the human body, hormones have been the most consistently exploited by junk-science entrepreneurs.

Before we discuss hormone junk science and aging, let's review a few facts.

The term "hormone" was first used by the physiologists W. Bayliss and E. Starling in 1902 to describe the action of secretin, a compound produced by the duodenum that stimulates the secretion of pancreatic juice into the gastrointestinal tract to aid digestion. Hormones are signaling molecules, and a vast number of biologically active compounds can be called "hormones," but the usage in human physiology is usually restricted to signaling molecules secreted by glandular cells into the internal fluid compartments of the body, most frequently into the blood. In humans, the various hormones comprise an interconnected signaling network, with some hormones controlling the secretion of other hormones that in turn affect still other hormones. If the nervous system can be thought of as a rapid electrical signaling system, the hormone system is a slow chemical signaling system.

Hormones are usually found in blood in extremely small concentrations, the blood acting as a hormone conveyor to various body parts, and the hormones then interact with target cells in various tissues. Target cells, in turn, which may contain specific receptors for specific hormones, respond according to their degree of specialization and their functional and nutritional status. The target cell response is usually a complicated biochemical integration of hormonal and neural stimuli. Increased secretion of the powerful hormone adrenaline (epinephrine), for example, may result in the activation of organs and muscle systems to increase heart rate, blood pressure, and respiration in a time of stress.

What is called "growth hormone" (also known as somatotropin or somatotropic hormone) is in humans secreted by the anterior part of the pituitary gland, and its main effect is to stimulate growth. Human growth hormone (HGH) is mostly biosynthesized as the precursor of a single-chain protein consisting of 191 amino acids and with a molecular weight of 22,000. The genes encoding human growth hormone, prolactin hormone, and choriomammotropin hormone form a gene family with many identical sequences of nucleotide bases (high nucleotide homology).

HGH secretion usually occurs in bursts, especially during fasting, with cycles of approximately four hours duration. Secretion of HGH is stimulated by fasting, hypoglycemia (low blood sugar), the hormone glucagon, various estrogens, and the hormone arginine vasopressin, and enhanced during the first phase of sleep. Secretion is inhibited by hyperglycemia (high blood sugar), the hormone somatostatin, the insulin-like growth factor IGF-I, the glucocorticoids that influence metabolism, and GHRH (a growth-hormone releasing factor). HGH is secreted from glandular cell storage granules, and in vivo the hormone tends to promote synthesis of protein, promote reduced fat deposition, and promote growth and mitosis of cartilage cells. Deficiency leads to various types of pituitary dwarfism. Excess secretion, usually the result of a pituitary tumor, produces the disease syndromes gigantism, acromegaly, organomegaly, and various metabolic disorders.

Insulin-like growth factors (IGFs, or somatomedins) are peptides secreted in response to HGH by liver, muscle, cartilage, and bone. Besides stimulating protein synthesis, IGFs decrease protein breakdown (protein catabolism), stimulate release of glucose by the liver, and have other effects on metabolism.

ON JULY 5, 1990, a paper that appeared in *The New England Journal of Medicine* became the origin of a junk-science industry that is now in its second decade and still thriving. The paper had ten authors, the lead author D. Rudman, the authors affiliated with various medical installations in the Chicago and Milwaukee areas. The title of the paper was "Effects of Human Growth Hormone in Men over 60 Years Old."

In this study, twelve men older than sixty years of age who had low plasma levels of insulin-like growth factor I (an accepted marker for HGH plasma concentration) were treated with injections of human growth hormone three times weekly for six months. Results demonstrated that after six months these men had statistically significant increases in lean body mass and bone mineral, unlike a group of nine similar men who received no treatment.

Almost immediately upon publication of this study, antiaging entrepreneurs in and out of the medical profession recognized an opportunity to make heaps of money. HGH must be administered by injection, since as a polypeptide it will not survive the gastrointestinal tract after oral administration. Some American physicians were willing to give such injections "off-label" to healthy adults, but the most lucrative operations were set up in Mexico, where in special HGH clinics an extended injection series could be obtained for approximately $1,000 a month ($3,000 at present). Meanwhile, inside the United States, to provide a lower-cost domestic alternative to an enormous potential market, various "dietary supplement" providers began selling a repertoire of HGH "releasers," chemicals touted to stimulate the production of HGH by the pituitary gland. There has never been any demonstration that these releasers actually provoke HGH secretion, and since dietary supplements are no longer regulated by the U.S. Food and Drug Administration, no one is certain these releasers are not harmful. There are no data concerning efficacy and no data concerning side effects of the HGH releasers. Harmful effects of such supplements are not reported to any agency.

As for HGH injections as antiaging treatments, commenting about the Rudman et al. 1990 HGH study, a recent editorial in the same journal states: "Although the findings of the study were biologically interesting, the duration of treatment was so short that side effects were unlikely to have emerged, and it was clear that the results were not sufficient to serve as a basis for treatment recommendations."

And in the same issue of the journal, endocrinologist Mary Lee Vance points out that it's not known whether long-term administration of growth hormone in the elderly is potentially harmful—particularly with regard to the risk of cancer, given that older age is associated with an increased incidence of cancer. In 152 healthy men in another study, the relative risk of the subsequent development of prostate cancer was increased by a factor of 4.3 among men who had serum concentrations of insulin-like growth factor I in the highest quartile, as compared with those whose concentrations were in the lowest quartile. This finding does not demonstrate causality by growth hormone or insulin-like growth factor I, but it does raise concern about giving older men growth hormone, which increases serum concentrations of insulin-like growth factor I. As Mary Lee Vance says, "There is no current 'magic bullet' medication that retards or reverses aging."

The situation concerning human growth hormone is clear: Administration of human growth hormone to elderly adults as an "antiaging" treatment based on a single unconfirmed study is junk science and is potentially dangerous. The sale of so-called HGH releasers is also based on junk science, also potentially dangerous, and just one example of public harm resulting from deregulation of the dietary supplement industry.

LONGEVITY THROUGH MEDITATION

INSTEAD OF OFFERING PILLS, SOME antiaging schemes propose meditation as a path to longevity. Consider the following description of the supposed antiaging benefits of meditation:

Although associated with the hippy culture, particularly in the past, there is now strong evidence that meditation promotes several health benefits including antiaging. Meditation carried out on a regular basis slows down the aging process by decreasing the amount of thyroxin produced by the thyroid gland, which in turn allows the body's processes to slow down to optimal levels and keeps the thyroid in balance. Meditation also stimulates the pineal gland at the center of the brain, the gland acting as an off switch to the many hormones that contribute to aging.

There is no specific public danger here; there is no evidence that meditation ever harmed or killed anyone attempting life extension. But the above paragraph demonstrates the junk-science attitude typical of many antiaging entrepreneurs. There is no scientific evidence that meditation promotes antiaging. There is no scientific evidence that decreasing human thyroxin production by the thyroid gland slows the aging process. There is no scientific evidence that meditation stimulates the human pineal gland, or that the human pineal gland acts as "an off switch" to many hormones. The pineal gland may release the hormone melatonin, which in turn apparently blocks the release of luteinizing hormone, but in general very little is known about the functions of the pineal gland, and the quoted paragraph is blatant junk science.

SCIENCE AND THE AGING PROCESS

So where are we? Is there any evidence that anything at all retards the aging process in humans? Beyond a healthy diet and general health, the answer is no: No magic antiaging bullet, no magic antiaging repertoire of pills, no magic injections, no magic creams or sprays, no schemes cheap or costly. But that does not mean there is no active scientific research underway on longevity in man and other animals. In the past several decades, research in the biology of aging has blossomed.

One of the central questions, perhaps the most important question directly related to human longevity, is simple in statement: What determines the life spans of animals? We can measure the average life spans of various species with exactitude, from the most primitive to the most complex, but what determines the numbers?

Medical geneticist Aubrey Milunsky discusses the variation among familiar animals and points out that life span of the mouse rarely exceeds two years; that of rats, four years; of cats, thirty years; of elephants, sixty years; and of horses, forty years. Bats, with an average life span of about thirty years, age much more slowly than mice, although the two species have similar body weights and cell chemistry (metabolic rates). Humans and virtually all animals age. Exceptions are those species that continue to grow in size after reaching adulthood. For example, several fish and reptile species show no biological age changes. They are not, however, immortal: They inevitably succumb to disease, predators, or accidents.

Developmental biologist Scott Gilbert points out that life expectancy, the amount of time a member of a species can expect to live, is not a characteristic of species, it's a characteristic of populations. Life expectancy is usually defined as the age at which half the population still survives. A baby born in England in the 1780s could expect to live to be thirty-five years old. In Massachusetts during that same time, the life expectancy was twenty-eight years. This was the normal range of human life expectancy for most of the human race during much of history. Even today, the life expectancy in some areas of the world (Cambodia, Togo, Afghanistan, and several other countries) is less than forty years. In the United States today, a child born in 1986 can expect to live seventy-one years if male and seventy-eight years if female.

Most multicellular organisms exhibit a progressive and irreversible physiological decline that characterizes what is called "senescence"—the aging process. The molecular basis of this process is unknown, but various mechanisms have been postulated, including the following: cumulative damage to DNA leading to genome instability; biochemical pathway alterations that lead to changes in gene expression patterns; chromosome telomere shortening in replicative cells; oxidative damage to critical macromolecules by reactive oxygen species; and nonenzymatic glycation of proteins.

Nematode worms are one of the most important laboratory animals in aging research. These worms are an abundant and ubiquitous phylum of unsegmented roundworms. *Caenorhabditis elegans* is a small (1 mm) nematode, transparent, hermaphroditic, free-living, and found in soil. It has a relatively small genome (approximately 19,000 genes) and only a few types of cells in its body. It has a sixteen-hour embryogenesis that can be achieved in a petri dish and is thus highly suitable for the study of developmental and behavioral genetics. Since nematodes are numerically so abundant—millions of individuals per square meter of soil—the worm *C. elegans* represents a diverse and successful group of animals. The simplicity of C. elegans, its transparency, and the ease with which it can be maintained in the laboratory have made it vital to the understanding of gene function. This tiny worm has also become famous as an example of how laboratory manipulations of genes can change the life span of an animal.

Experimental genetic manipulation of the aging process in multicellular organisms has been achieved in the nematode worm C. elegans through alterations in the insulin receptor pathway, in the fruit fly Drosophila through the overexpression of certain enzymes, and in both organisms through the experimental selection of stress-resistant mutants. In mammals, however, the only intervention that appears to slow the intrinsic rate of aging is caloric restriction.

Most studies of caloric restriction in mammals have involved laboratory rodents subjected to a long-term 25 to 50 percent reduction in caloric intake without deficiency of essential nutrients, and the result in these rodents is a delayed onset of age-associated pathological and physiological changes and an extension of maximum life span. Various mechanisms have been postulated to explain this result, including increased DNA repair capacity, altered gene expression, depressed metabolic rate, and reduced oxidative stress.

Another approach views aging in terms of natural selection of functions and connections between functions (functional pathways). Researchers M. R. Rose and A. D. Long point out that most biological research attempts to uncover functional pathways, whether specific biochemical reactions or large-scale developmental processes. Such pathways evolve by natural selection. Mutation and other genetic tricks that modify functional pathways usually impair their operation. But aging is not a function. Aging can be defined as an endogenous progressive deterioration in age-specific components of fitness. Aging is not a target of evolution by natural selection. Evolution does not select long-lived animals for survival. Aging is instead a secondary effect of the decline in the impact of natural selection with age. From such a theory it follows that many loci, and many biochemical pathways, are expected to produce the damaging effects of aging—because aging is a secondary side effect of normal evolution.

What can be said about the genes involved in aging? In C. elegans, yeast (C. cerevisiae), the fruit fly Drosophila melanogaster, and even in mice, there are many genes that upon mutation increase longevity. In many cases, the proteins encoded by such genes have equivalents in higher animals. For example, the adult life span of C. elegans can be markedly increased as a result of reduced activity of a signaling pathway resembling that which responds to insulin or insulin-like growth factor-I (IGF-I) in humans. But the important question is whether such genes control aging in all animals. Has the biology of aging been conserved throughout evolution? Most types of animals grow old and die, but it does not follow that aging involves the same processes or the same genes in all species. It's indeed possible that in different species the underlying mechanisms of aging are different.

HAYFLICK AND CELLULAR SENESCENCE

CELLULAR SENESCENCE, THE MORTALITY OF individual cells, is also thought to contribute to aging, although how it does so is poorly understood. In addition to arrested growth, senescent cells show changes in function, and since they accumulate with age, they may contribute to age-related declines in tissue and organ physiology.

Although cellular senescence may be of great importance in understanding the aging process, there's a tendency in aging research to ignore cellular senescence as a fundamental factor and focus instead on diseases of old age.

Leonard Hayflick, a major researcher in cell senescence, points out that in the past hundred years life expectancy at birth in developed countries has increased from approximately forty-eight years to seventy-six years, the same gain that occurred over the previous 1,900 years. But this progress has neither advanced nor resulted from our understanding of aging. Instead, it's the control of infectious diseases of the young that explains the increase in life expectancy during the twentieth century. Hayflick notes that the failure to distinguish between the diseases of old age and the aging process is widespread even in the scientific community. The virtual resolution of various childhood diseases such as poliomyelitis and iron-deficiency anemia did not increase our knowledge of childhood development. Similarly, the resolution of the leading causes of death in old age—cardiovascular disease, stroke, and cancer—are unlikely to advance our knowledge of the aging process.

According to Hayflick, one example of the consequences for science policy of the failure to distinguish research on age-associated diseases from research on the fundamental biology of aging is that "it's virtually impossible to raise funds for research on aging, because in the minds of policymakers and the public no one suffers or dies from it." More than half of the budget of the U.S. National Institute on Aging is spent on Alzheimer's disease, yet the elimination of this disease "will have only a trivial impact on life expectancy and will not advance our knowledge of the fundamental biology of aging." Hayflick suggests that greater attention must be given to a question that is rarely posed: Why are old cells more vulnerable to disease than young cells? "The resolution of all causes of death currently written on the death certificates of those older than sixty-five will result in an increase in life expectancy of only about fifteen years. An increase in our knowledge of how age changes occur does not put a fifteen-year limit on what is possible."

AGING AND TELOMERES

"TELOMERES" ARE DEFINED ENDS OF chromosomes that contain specific repeated DNA sequences apparently essential for normal chromosome replication. Since their length shortens a bit with each replication, they may be involved in the aging of the cell.

Over sixty years ago Barbara McClintock (1902–1992) described the telomere and suggested that it protected the chromosome from deleterious end-to-end fusion, thus functioning to protect the genome. Since that time researchers have discovered that the telomere is a complex structure composed of both DNA and a growing list of associated proteins that together serve to regulate the length of the telomere and protect genomic integrity. In addition to its protective role, the telomere has also been hypothesized to serve as a molecular clock that tallies the number of cell divisions and limits further divisions at a predetermined point. However, the precise role of telomeres in predicting and limiting cellular life span remains a matter of much debate.

Insight accumulates through research. The enzyme telomerase, the reverse transcriptase that synthesizes telomeric repeats, is present at low levels in human tissue stem cells, progenitor cells, and germ cells, and is undetectable in the vast majority of adult somatic tissues. Insufficient telomerase activity and the inability of DNA polymerase to replicate the extreme ends of chromosomes lead to telomere attrition with each round of cell division in the context of organ renewal during advancing age, high-turnover disease states, and replication in tissue culture.

In humans, progressive shortening of telomeres provokes replicative senescence after sixty to eighty divisions of primary human fibroblasts (the main cells of connective tissue) in tissue culture. Introduction of telomerase into such cells stabilizes telomeres, prevents senescence, and endows cells with unlimited proliferative potential. As for cancer, the ability of telomerase to rescue cells from the adverse consequences of telomere dysfunction is believed by some to be critical for the role of telomerase in facilitating malignant transformation of primary human cancer cells and in maintaining the viability of established cancer cells.

AGING AND DNA DAMAGE

ONE OF THE FACTORS POSTULATED to drive the aging process is the accumulation of DNA damage. Researchers have provided strong support for this hypothesis by studies of mice with a mutation in XPD, a gene encoding a specific enzyme (a DNA helicase) that functions in both repair and transcription and that is mutated in the human disorder TTD. The term "trichothiodystrophy" (TTD) refers to a human genetic defect that produces brittle hair from low content in hair of the sulfur-containing amino acid cysteine. Mice with TTD exhibit many symptoms of premature aging, including osteoporosis and a forward curvature of the spine (kyphosis), abnormal hardening of bone (osteosclerosis), early graying, general weight loss and wasting (cachexia), infertility, and reduced life span. TTD mice carrying an additional mutation in the gene XPA, a mutation that worsens the DNA repair defect, show a greatly accelerated aging phenotype that correlates with an increased cellular sensitivity to oxidative DNA damage. Researchers believe a reasonable hypothesize is that aging in TTD mice is caused by unrepaired DNA damage that compromises transcription, leading to functional inactivation of critical genes and increases in cell suicide (apoptosis).

SUCH IS THE CURRENT science of the aging process. It's evident that we're far from understanding why aging occurs, either in cells or in tissues or in whole people. It's also evident that hawkers of schemes to prevent aging and prolong longevity can't possibly have schemes based on science because the science of aging is at present too meager. That won't stop the hawking, of course, but the public needs to be aware that most of the hawking involves junk science.

BIO-LUDDITES AGAINST LONGEVITY

IT'S AN IRONY THAT WHILE so much junk science is devoted to selling the idea of increasing longevity, a parallel junk science proposes reasons to avoid longevity. These days the opposition to extending life is primarily an opposition to biological science and biotechnology—the opposition of so-called bio-Luddites, a bio-Luddite pessimism versus futurist optimism.

Within the bio-Luddite group there are factions opposed to each other in everything else. In one bio-Luddite camp are the classical Luddites who fear science and technology in any form, calling them weapons of the establishment. In the other bio-Luddite camp are neoconservatives of the establishment itself, a coterie of academic, media, and political people who fear attempts to extend human life span because they view such extension as a potential threat to the established order. They may be right, but the threat is long-term, not near-term, not a twenty-first century threat at all, maybe not even a third millennium threat. Aging and life extension are active fields of biological research, but like most current fields in biology, the advent of molecular biology and the introduction of new technologies has forced things to begin afresh. Aging research is in fact still in its beginning, and the idea that the human species is on the verge of immortality is without scientific basis.

MEANWHILE ENTREPRENEURS CONTINUE TO prey on public vulnerabilities with myths concerning human aging and life extension, selling potentially dangerous pills, potions, and nostrums to the gullible and desperate public. Was the recent deregulation of dietary supplements a good idea? It may have been good for various entrepreneurs, but not at all for the American public.

6

TOBACCO: DRUG DEALING IN AMERICA

I believe that nicotine is not addictive.
— Thomas E. Sandefur, Jr., Brown & Williamson
Tobacco Corporation, 1997

The Brown & Williamson documents revealed that both Brown & Williamson and British American Tobacco Company clearly recognized that their own sponsored research showed that nicotine is pharmacologically active, that it is addictive, and that cigarettes are in essence nicotine delivery devices.
— R. S. Feldman et al., *Principles of Neuropharmacology*, 1997

WE CAN EXPECT THAT centuries from now the story of the tobacco industry in the twentieth and twenty-first centuries will be taught as a case study in the abuse of public trust by corporate profiteering. The scale is so vast, the calumny so blatant, the twisting of science so insidious, the misery and death produced so overwhelming, any account cannot possibly do the realities full justice.

Although cigarettes were already common during the American Civil War, nearly a hundred years passed before the dangers of cigarette smoking were publicly recognized — and vigorously denied by tobacco companies.

According to the numbers that are available, between 1950 and the year 2000, it is estimated that American and British tobacco companies may have killed more people than all the wars and genocides of the twentieth century combined. The executives of these companies accomplished this remarkable feat by packaging an addictive drug in a device whose use causes lethal disease: cancer, heart disease, stroke, and emphysema. Once addicted, many people smoked cigarettes until their dying breath.

Are the numbers really clear? Yes, they are. Tobacco use is still the leading preventable cause of death in the United States, responsible for more than 400,000 deaths annually, or one of every five deaths. Half of regular smokers die prematurely of a tobacco-related disease. What these statistics mean is that during the past fifty years approximately 20 million people in the United States alone have died as a result of tobacco use. Worldwide deaths are estimated to be currently 2.5 to 5 million per year, which means in the past twenty years alone 50 to 100 million people have been killed by tobacco smoking.

Tobacco is now a global problem. Government officials around the world now recognize what industry executives have long understood—the tobacco business is fundamentally a global enterprise. The sale of raw leaf and finished products, the smuggling of cigarettes to evade taxes, and the effects of media advertising all cross national borders. The consequences of this enterprise are staggering—by the year 2020, an estimated 8.4 million people will die annually from tobacco-related diseases, more than two-thirds of them in developing countries, where tobacco companies are now mounting intensive advertising campaigns to sell addictive cigarettes. If current trends continue, more people will perish annually from tobacco-related illness than from any single disease.

"Big Tobacco" refers to the five companies that produce 97 percent of the cigarettes in the United States. The largest company is Philip Morris, originally a British firm that entered the American market in 1902. During the past fifty years, Philip Morris has acquired the largest share of the American market, and in the world market is second only to the China National Cigarette Company. Marlboro is a Philip Morris brand. The second largest company in the United States, British American Tobacco, owns Brown & Williamson. The third largest company in the United States, R. J. Reynolds (merged with Brown & Williamson in July 2004), manufactures Camel, Winston, and Salem cigarettes. The fourth and smaller company is Lorillard, maker of Newport cigarettes. The smallest of the big five is Liggett Tobacco (formerly part of the Liggett Group and now part of the Brooke Group).

THE RECORD OF THE TOBACCO INDUSTRY

THE TOBACCO STORY, AS A case illustrating the use of junk science to deceive the public, needs a historical perspective. Until the early 1950s, it was not possible to analyze in detail the contents of tobacco smoke because the scientific instrumentation was not available to identify the various carcinogens. Similarly, it was only in the late 1950s that the first conclusive animal experiments were completed indicating that tobacco smoke causes lung tumors. Most of this research was done by the tobacco industry itself, until finally the research efforts were shut down by tobacco-industry attorneys who feared the results of the research might leak to the public and open the tobacco industry to litigation. By the end of the 1950s, the causal link between tobacco smoke and lung cancer and the addictive properties of nicotine were well-known to the tobacco industry and beginning to be known to the medical community and the public. What is clear from the record is that from the 1950s to the present day, the overall strategy of the tobacco industry has been to twist science in public statements in order to avoid litigation and regulation—a carefully designed strategy to maintain profits at the expense of public health.

Here are some realities coupled with some of the junk-science public statements by tobacco-industry executives through the years.

Researchers reported in 1950 that 96.5 percent of lung-cancer patients are moderate to heavy smokers.

Researchers reported in 1953 that painting cigarette tar on the backs of mice creates tumors.

UK Tobacco Institute Research Council, 1954: "Distinguished authorities point out: that medical research of recent years indicates many possible causes of lung cancer; that there is no agreement among the authorities regarding what the cause is; that there is no proof that cigarette smoking is one of the causes; that statistics purporting to link smoking with the disease could apply with equal force to any one of many other aspects of modern life. Indeed the validity of the statistics themselves are questioned by numerous scientists."

R. J. Reynolds Tobacco Company, 1954: "There still isn't a single shred of substantial evidence to link cigarette smoking and lung cancer directly."

UK Imperial Tobacco, 1956: "I state that in our considered opinion there is no proof at all that smoking causes lung cancer and much to suggest that it cannot be the cause."

In 1963, Addison Yeaman, general counsel to the Brown & Williamson Tobacco Corporation, wrote in a corporate memorandum: "Nicotine is addictive. We are, then, in the business of selling nicotine, an addictive drug."

In 1964, U.S. surgeon general Luther Terry issued a general report citing health risks associated with smoking.

Philip Morris, 1964: Following the U.S. surgeon general's report of January 1964, a Philip Morris director dismissed the findings: "We don't accept the idea that there are harmful agents in tobacco."

Philip Morris, 1968: "No case against cigarette smoking has ever been made despite millions spent on research. . . . The longer these tests go on, the better our case becomes."

Brown & Williamson Tobacco Corporation, 1971: "It is our opinion that the repeated assertion without conclusive proof that cigarettes cause disease—however well-intentioned—constitutes a disservice to the public."

UK Imperial Tobacco, 1975: "As a company we do not make, indeed we are not qualified to make, medical judgments. We are therefore not in a position either to accept or to reject statements made by the Minister of Health."

Philip Morris, 1976: "None of the things which have been found in tobacco smoke are at concentrations which can be considered harmful. Anything can be considered harmful. Apple sauce is harmful if you get too much of it."

In 1982, U.S. surgeon general C. Everett Koop reported that second-hand smoke may cause lung cancer.

Tobacco Institute of Hong Kong, 1989: "The view that smoking causes specific diseases remains an opinion or a judgment, and not an established scientific fact."

UK Tobacco Institute, 1998: Murray Walker, vice president and chief spokesperson for the Tobacco Institute, testifying at the Minnesota trial, the trial of the states attorneys general against the major tobacco companies: "We don't believe it's ever been established that smoking is the cause of disease."

Philip Morris, 1998: Geoffrey Bible, chairman of Philip Morris, testifies at the Minnesota trial: "I'm unclear in my own mind whether anyone dies of cigarette smoking–related diseases."

UK Tobacco Manufacturers Association, 1998: John Carlisle of the UK Tobacco Manufacturers Association is questioned in a magazine. Question: "Does it [smoking] cause lung cancer?" John Carlisle: "There's no shortage of statistics: It's extraordinary the amount of research that has gone into our product and the many and varied opinions that people hold about it."

Finally, R. J. Reynolds Tobacco Company, 1992, a private statement: An actor promoting R. J. Reynolds products evidently asked an RJR executive why the executive does not smoke. The actor is told: "We don't smoke that s***. We just sell it. We just reserve the right to smoke for the young, the poor, the black, and the stupid."

TOBACCO TOXICITY

ONE OF THE IRONIES OF contemporary society is that toxic chemicals that enter the body through the digestive tract are more tightly regulated by governments than toxic chemicals that enter the body through the lungs. If the plant called "lettuce" were found to cause millions of cases of cancer, stroke, heart disease, and emphysema, lettuce would quickly vanish as a marketable commodity, no matter how many lettuce growers squealed in protest. But the plant tobacco, which does cause millions of cases of cancer, stroke, heart disease, and lung disease, is not only still a marketable commodity in the United States and elsewhere, but more young people than ever in the United States (in both numbers and percentages), and more people worldwide, are smoking cigarettes. In the United States, a contagious disease that caused as many deaths as cigarette-selling would be called a plague; instead, in the United States, cigarette-selling is called an "industry." But if cigarette-selling is an industry, it's an industry devoted to the sale of an addictive lethal drug, a drug no less lethal just because it kills people in twenty or more years rather than killing people immediately.

Tobacco smoke contains approximately 4,000 compounds. The most important carcinogens in tobacco smoke are polycyclic aromatic hydrocarbons, arylamines, and N-nitrosamines. Individuals vary in their ability to convert these toxic metabolites of cigarette smoke into less harmful compounds—which is the explanation for the fact that not everyone who smokes contracts a serious disease.

The addictive component in cigarettes is nicotine, a chemical well-known to pharmacologists as a poison. Nicotine is one of several alkaloids that can be extracted from tobacco leaves, constituting approximately 5 percent of the total weight of the dry plant leaves. Inhaled tobacco smoke contains a complex mixture of substances, including carbon monoxide and various carcinogenic hydrocarbons generated by the tobacco combustion process in the smoked cigarette. These inhalants constitute the tobacco "tar" that provides the primary taste and smell of the smoke. The tar and carbon monoxide are responsible for most of the diseases associated with long-term tobacco use.

One important effect of cigarette smoking is on the unborn fetus. In the

United States, 65 percent of all infant deaths occur among low-birth-weight infants (less than 2,500 grams), with such infants accounting for 7.6 percent of all live-born infants. The causes of low birth weight are largely unknown, but both environmental and genetic factors may play a role. Numerous studies have demonstrated that maternal cigarette smoking during pregnancy is associated with reduced birth weight or increased risk of low birth weight. In 1997, thanks to tobacco industry advertising and little if any counteractive advertising, as much as 13.2 percent of U.S. women reported smoking cigarettes during pregnancy. Maternal cigarette smoking is identified as the single largest modifiable risk factor for intrauterine growth restriction in developed countries. Where are the public service billboards cautioning pregnant women not to smoke cigarettes?

In 1998 in the United States, an estimated 47 million adults (24 percent of the adult population), comprising 25 million men (26 percent) and 22 million women (22 percent) were current smokers. Overall, 20 percent of adults were everyday smokers, and 4 percent were some-day smokers. Everyday smokers constituted 82 percent of all smokers. Prevalence of smoking was highest among persons aged eighteen to twenty-four years (28 percent) and aged twenty-five to forty-four years (27 percent), and lowest among persons aged greater than sixty-five years (11 percent). Current smoking prevalence was lowest among persons with at least sixteen years of education (11 percent) and highest among persons with nine to eleven years of education (37 percent). Smoking prevalence was higher among persons living below the poverty level (32 percent) than among those living at or above the poverty level (23 percent). These data clearly indicate that the most vulnerable people in society, the young, the uneducated, and the poor, are the people most affected by tobacco-industry advertising.

Cigarette smoke is so toxic, there is even danger for people near the people who smoke. The dangers of "passive smoking" have been hotly debated in and out of restaurants and bars for decades, but the evidence is as clear as that for active smoking. Passive smoking has been identified as an important risk factor for cardiovascular disease. In 1992, the American Heart Association concluded that the risk of death due to heart disease is increased by approximately 30 percent among those exposed to environmental tobacco smoke at home, and could be much higher in those exposed at the workplace, where higher levels of environmental tobacco smoke may be present. There is evidence that exposure of nonsmokers to environmental tobacco smoke breaks down serum antioxidant defenses and is associated with damage to arteries (impairment of endothelium-dependent function of arterial walls). Coronary flow velocity reserve (CFVR) is a measure of endothelial function in the coronary circulation. Recent research indicates that "passive smoking substantially reduced CFVR in healthy nonsmokers. This finding provides direct evidence that passive smoking may in some people cause endothelial dysfunction of the coronary circulation in nonsmokers."

PUT ASIDE FOR A moment the diseases caused by cigarette smoke and focus on the essential aspect of cigarettes: They are delivery vehicles for an addictive drug, and all the claims of the tobacco industry to the contrary are blatant junk science. The major psychoactive ingredient in tobacco is nicotine, 1-methyl-2-(3-pyridyl) pyrrolidine. A typical American cigarette contains approximately 9 milligrams of nicotine, with a yield to the smoker of approximately 1 milligram. When tobacco smoke is inhaled, nicotine readily passes through the absorbent surfaces of the lungs. When tobacco is chewed or snorted, nicotine is absorbed to a lesser degree through the tissues of the mouth. When tobacco smoke is inhaled, 25 percent of the nicotine reaches the brain in approximately seven seconds, twice as fast as when the drug is administered intravenously. Thus, for the addictive agent nicotine, tobacco smoke inhalation via the modern cigarette is the fastest and most efficient method of delivery of this drug to the human brain.

Nicotine is a well-known pharmacological agent. Systemic doses of the nicotine absorbed during tobacco smoking exert profound pharmacological effects on the human central nervous system. Circulating nicotine readily passes the blood-brain barrier to enter the brain and activate nicotinic cholinergic receptors. In addition to affecting motivational and cognitive processes, nicotine can produce tremors and convulsions at high doses. At lower doses, nicotine increases respiration rate, acts on vasomotor centers in the medulla to induce cardiovascular changes, and causes the release of antidiuretic hormone from the pituitary gland.

The current consensus in the medical and pharmacological communities is that nicotine in tobacco is an addictive drug. Abrupt abstinence from regular tobacco use can result in withdrawal symptoms severe enough to make abstinence extremely uncomfortable. The abstinence symptoms consist of severe craving, irritability, anxiety, difficulty concentrating, restlessness, decreased heart rate, dysphoria, impatience, insomnia, and increased appetite and weight gain. A return to smoking removes the symptoms and increases the reinforcing properties of the drug.

In addition to its addictive properties, nicotine is highly toxic, with toxic doses producing the following consequences: excitement, confusion, muscular twitching, weakness, abdominal cramps, clonic convulsions, depression, rapid respiration, palpitations, collapse, coma, central nervous system paralysis, and respiratory failure.

Much has been written about the sociopsychological factors involved in cigarette smoking, factors such as social class and personality profile. But the important aspect is the physiological addiction: The rewarding effects of tobacco use and the craving associated with withdrawal reflect a true chemical dependence. Cigarette smoking is not a "habit," it's an addiction to a drug and the drug is nicotine. Recent research has clarified the molecular interactions of nicotine with neuron receptors and related these interactions to nicotine addiction.

In public, however, the tobacco industry continues to hide its own scientific evidence. The tobacco industry would like people to believe that the craving, when it exists, is for the cigarette, the taste, the slim white tube, the allure of smoking carefully cultivated by advertising. They will not admit in public that the nicotine in cigarettes is and has been an addictive drug. Here is James Morgan, president of Philip Morris:

> If [cigarettes] are behaviorally addictive or habit forming, they are much more like . . . Gummi Bears. I love Gummi Bears . . . and I want Gummi Bears, and I eat Gummi Bears, and I don't like it when I don't eat my Gummi Bears, but I'm certainly not addicted to them.

CIGARETTES AS DRUG DELIVERY DEVICES

THE TOBACCO-INDUSTRY LITIGATION DOCUMENTS REVEAL that for decades the tobacco industry knew and internally acknowledged among themselves that nicotine is an addictive drug and that cigarettes are the ultimate nicotine delivery device. The following internal company statement by executives is found in tobacco-industry documents: "Very few consumers are aware of the effects of nicotine, i.e., its addictive nature and that nicotine is a poison." (H. D. Steele, Brown & Williamson Tobacco Corporation, 1978).

And in another Brown & Williamson memo: "Nicotine is the addicting agent in cigarettes." (A. J. Mellman, Brown & Williamson Tobacco Corporation, 1983).

Concerning cigarettes as a drug delivery device, litigation documents reveal that C. E. Teague, Jr., assistant director of research at R. J. Reynolds Tobacco Company, wrote in 1972 in an internal memorandum: "In a sense, the tobacco industry may be thought of as being a specialized, highly ritualized and stylized segment of the pharmaceutical industry. Tobacco products, uniquely, contain and deliver nicotine, a potent drug with a variety of physiological effects. . . . Thus a tobacco product is, in essence, a vehicle for delivery of nicotine."

Thus, inside the industry the nicotine in cigarettes is known through research as an addictive drug and cigarettes are known as vehicles for that drug, while outside the industry cigarettes are presented to the public as mere desirable habit-formers no more addictive than Gummi Bears. The blatant corruption of science is startling.

There is even solid evidence of tobacco-industry efforts spanning three decades to alter the chemical form of nicotine to increase the percentage of freebase nicotine delivered to smokers. The chemical form of nicotine depends on the acidity of its surroundings. Depending on pH, nicotine exists as a diprotonated salt, a monoprotonated salt, or an uncharged neutral species. The salt forms are called the "bound" forms, and the neutral species is called the "freebase" form. Nicotine favors the salt form at acidic low values of pH (e.g., pH = 3) and the freebase form at alkaline high values of pH (e.g., pH = 8). Freebase nicotine apparently crosses biological membranes more easily than the charged counterparts, and this affects the physiological response to the drug. These tobacco-industry manipulations of the chemistry of cigarettes to make cigarette nicotine more potent and addiction more powerful were never revealed to the public.

No matter the repeated claims of tobacco-industry executives, the science is clear: Nicotine is a highly toxic drug with an adult human lethal dose of only 60 milligrams. At sublethal doses, nicotine acts on the central nervous system to affect motivational and cognitive neural processes, and to produce psychiatric and physiological symptoms associated with drug addiction. The drug dealing and junk-science dealing by the tobacco industry is an insidious public menace because the delivery vehicles for the addictive drug nicotine, the cigarettes sold to be smoked, are responsible for an array of fatal and debilitating diseases, including lung cancer, stroke, and cardiovascular disease.

TARGETING YOUNG PEOPLE

THERE IS EVIDENCE THAT CIGARETTE companies may have targeted young people in their magazine advertising and that cigarette advertising in magazines is likely to reach a substantial number of young people. In November 1998, the attorneys general of forty-six states signed a "Master Settlement Agreement" with the four largest tobacco companies in the United States. The agreement states that cigarette companies may not "take any action, directly or indirectly, to target youth . . . in the advertising, promotion or marketing of tobacco products." In June 2000, Philip Morris announced that beginning in September 2000 it would restrict its cigarette advertising to magazines whose proportion of young readers was less than 15 percent and that had fewer than 2 million readers from twelve to seventeen years old. R. J. Reynolds declined to adopt a similar policy.

Researchers have analyzed trends in expenditures for advertising for fifteen specific brands of cigarettes and the exposure of young people to cigarette advertising in thirty-eight magazines between 1995 and 2000. In 2000 dollars, the overall advertising expenditures for the fifteen brands of cigarettes in the thirty-eight magazines were $238 million in 1995, $219 million in 1998, $291 million in 1999, and $217 million in 2000. Expenditures for youth brands (cigarette brands such as Joe Camel and Players Light designed to especially appeal to young people) in youth-oriented magazines were $56 million in 1995, $58 million in 1998, $67 million in 1999, and $60 million in 2000. Expenditures for adult brands in youth-oriented magazines were $72 million, $82 million, $109 million, and $68 million, respectively. In 2000, magazine advertisements for youth brands of cigarettes reached more than 80 percent of young people in the United States an average of seventeen times each. The researchers concluded: "The Master Settlement Agreement with the tobacco industry appears to have had little effect on cigarette advertising in magazines and on the exposure of young people to these advertisements."

Most analyses of tobacco-industry marketing have demonstrated that the tobacco industry targets youth. The tobacco industry also conducts extensive marketing research on adults, particularly young adults aged eighteen to twenty-four years. The industry's interest in young adults is not surprising, since young adults compose its youngest legal marketing target, vastly outnumber teenage smokers, are making the transition from smoking initiation to becoming fully addicted smokers, and are role models for teenagers. Tobacco companies have used detailed studies of young adult smokers' motivations, aspirations, activities, and environment to guide cigarette marketing and advertising campaigns.

Psychological and sociological profiles of cigarette smokers are a subject of intense study by psychologists and sociologists—and especially by the tobacco industry, which gears its advertising according to such profiles. Mental illness and smoking have been closely linked. For example, smoking rates have been reported to be over 80 percent among persons who have schizophrenia, 50 to 60 percent among persons with depression, 55 to 80 percent among those who have alcoholism, and 50 to 66 percent among those who have substance-abuse problems. One study estimates that smokers with coexisting psychiatric or substance-abuse disorders account for 44 percent of all cigarettes smoked in the United States, a percentage that reflects both the high prevalence of smoking in connection with these conditions and the fact that patients with these disorders are very heavy smokers.

As drug dealers, the tobacco companies have been immensely successful, have accumulated an enormous amount of capital, and have been actively acquiring influence in other industries. The diversification of the tobacco industry is well-documented. The tobacco industry has systematically acquired companies that manufacture unrelated consumer products such as cookies, macaroni and cheese, candy, and pharmaceuticals, and the tobacco industry has used its financial ties to pressure a variety of industries to oppose tobacco control. The pharmaceutical industry also maintains diversified interests and is involved in the sale of multiple products such as chemicals, pesticides, plastics, and pharmaceuticals. Corporate diversification has therefore resulted in financial ties between pharmaceutical companies that market nicotine replacement therapies and the tobacco industry. One industry sells an addictive drug, an associated industry sells other drugs to overcome the addiction, and caught between these pincers is the poisoned public.

TOBACCO AND THE GOVERNMENT

WHAT DO WE DO? OVER the years, the U.S. government has exhibited marked ambivalence concerning what to do about the tobacco industry problem. In

1994, the Food and Drug Administration sought federal regulation of the content of tobacco products, but in the year 2000 this was rejected by the U.S. Supreme Court. Nicotine replacement therapy as a cure for nicotine addiction is regulated by the Food and Drug Administration, but nicotine-containing cigarettes are not regulated. Federal agencies concerned with agriculture and health are inherently in conflict over what to do about tobacco. There is no apparent anti-tobacco champion in the Bush administration, although the attorney general in the Bush administration is in charge of the current $300 billion case against the tobacco industry.

Warnings by the U.S. surgeon general on cigarette packages in the United States are tame by comparison with warnings in other countries. In Australia, every package of cigarettes carries a large warning: "Smoking Is Addictive." In Canada, cigarette packages carry the warning, "Cigarettes Cause Mouth Diseases," the warning overlaying a disturbing photograph of a mouth with unhealthy gums and teeth.

The difficulty even advanced countries have in simply banning the production and sale of cigarettes is a testament to the grip the tobacco industry has on these governments.

THROUGHOUT THE SECOND HALF of the twentieth century, the tobacco industry presented to the public and to legislators in the United States and the UK views about cigarette smoking that were based on junk science or no science at all or were blatant lies. The real scientific data about nicotine addiction and cancer obtained by the tobacco industry in its own research was kept secret and revealed only in the course of litigation toward the end of the century.

The corruption of reason and drug dealing endure. Deaths due to tobacco use occur primarily in middle age. Those who have lost brothers, sisters, wives, husbands, mothers, fathers, and friends to the machinations and junk science of the tobacco industry, apart from their grief, have reason to feel anger at a great public injustice. As for the millions most affected by the machinations and junk science of the tobacco industry, the question of anger is moot—they're all dead.

PART THREE

MEDICAL FOLLIES

7

JUNK MEDICINE, BIG PHARMA, BIG PROFITS

Cocaine Toothache Drops. Instantaneous Cure! Price 15 Cents. Prepared by the Lloyd Manufacturing Co. 219 Hudson Ave., Albany, N.Y. For Sale by All Druggists.
— ADVERTISEMENT, LATE NINETEENTH CENTURY

The pretense that pharmaceutical marketing is education requires the participation of at least two parties—the industry and the medical profession.
— MARCIA ANGELL, M.D., FORMER EDITOR-IN-CHIEF OF
The New England Journal of Medicine

MEDICINE IS NOT AN easy vocation, not easy on the intellect and not easy on the emotions. If there is anything like an essential cadre in ordinary human existence, it's the cadre of physicians throughout the world. Medical people are usually called to medicine early, and for the most part they are a special breed among us. The preparation for a medical career is a difficult grind, not the ultimate in intellectual challenge, but a taxing grind on the spirit. And who are they, these physicians? Their lives are often examples of quiet ironies: In towns and villages everywhere in the world, one can find children laughed at for their books and science and secret hopes of doctoring, children who grow up to save the lives of those who laughed at them. That's the essence of it: Physicians save lives.

And here we are at the beginning of the twenty-first century writing about junk medicine and profits and big drug companies. Once upon a time physicians visited houses in the night to doctor children when they were ill, the physicians worn-out graying men with their little black bags and soft voices, physicians who often waved away payment when the family was poor. Such was an age of selfless medicine, and to write about the ugly profit-oriented side of contemporary medicine is a sad charge. But not enough is written about it, at least not enough so far to produce necessary corrections, not enough to restore the grace of the profession. We're in an era when physicians themselves publicly castigate their profession for its selfish greed—a phenomenon in the history of medicine, and an indication that something is rotten in the way we live.

JUNK MEDICINE

First, let's draw a parallel. A research scientist in an advisory capacity, to government or industry, for example, who recommends a course of action purportedly based on his scientific knowledge, when he knows but hides the fact that there is no evidentiary scientific support for his recommendation, and that given the absence of data, his recommended course of action may be harmful—that scientist is practicing junk science.

Similarly, a physician, in an advisory capacity to a patient, who writes a prescription for a drug for that patient, a recommendation purportedly based on the physician's medical knowledge, when the physician knows but hides the fact that there is not sufficient evidentiary medical knowledge to justify the use of that drug in the case of this patient, and the physician is aware that because of the possibility of unknown side effects the drug may be harmful to the patient—that physician is practicing junk medicine.

The crux of one of the important problems of modern medicine is clear: Junk medicine, as described above, is rampant in modern medicine, just as useless and as potentially dangerous to the patient as tribal witch-doctoring, and in the large a public danger of the first magnitude. Physicians are too often mere devices to facilitate a commercial transaction between their patients and drug companies, and the drug companies are well aware of this and devote an enormous amount of energy, time, and money to capturing and using their physician sales-facilitators. The drug business has succeeded in thoroughly corrupting the practice of medicine. To pharmaceutical companies, physicians practicing clinical medicine are sales-facilitators essentially bought the way actors are bought to wear white coats in front of television cameras while pretending to be physicians.

A second problem concerns medical research, or more aptly biomedical research, since these days so much medical research is coupled to modern biology, to molecular biology and biochemistry. During the past twenty-five years, we have witnessed a dangerous corruption of biomedical research by private enterprise, particularly by the large pharmaceutical companies—"big pharma." Most observers agree that this corruption began in the United States with the passage of the Patent and Trademark Law Amendments Act of 1980, known as the Bayh-Dole Act (named after Senator Birch Bayh [D-Ind.] and Senator Robert Dole [R-Kans.]). By enabling universities and small businesses to patent discoveries emanating from research sponsored by the National Institutes of Health (NIH) and then grant exclusive licenses to drug companies, this act set in motion processes that completely altered research in biomedicine, brought large research universities into partnerships with biotechnology companies and pharmaceutical companies, and pushed a large and important area of modern academic science to focus more on profits than on public benefit. The goose has laid a pile of golden eggs, and the eggs may eventually kill us.

PHYSICIANS AS DRUG SALES-FACILITATORS

HOW DO THEY DO IT? How do drug companies recruit their physician sales-facilitators? The easy answer is that the drug companies are awash in money. Americans are currently spending $200 billion a year on prescription drugs. That's two-tenths of a trillion dollars. That's approximately $700 a year for every man, woman, and child in the country. For the average family of four, that's $2,800 a year in revenue to the drug companies. For patients on permanent prescriptions, the average cost for each drug is $1,500 a year. It makes no difference that some families have prescription drug insurance: What is out there is a huge machine shoveling cash from various parts of the economy, public and private, into the hands of the pharmaceutical industry. Anything that can be bought, the drug industry can afford to buy it: politicians, physicians, scientists, journalists, medical associations, scientific journals, scientific societies, add what you like to the list.

The drug industry currently spends on "sales and administration" an amount equivalent to $8,000 to $15,000 per U.S. physician per year. The industry sales force of "drug detailers" who meet individually with physicians to promote company products numbers nearly 90,000 in the United States alone—one salesperson for every five physicians. The average physician is talking to drug detailers at least four times a month.

That is only the beginning. A recent article in the *British Medical Journal* catalogued sixteen different methods used by drug companies to directly or indirectly personally interact with physicians, methods ranging from dispensing of trivial gifts to ghostwriting articles for academic physicians, payment of large "honoraria" and consulting fees to prominent physicians who promote company products, and lavish travel and entertainment for physicians who actively prescribe the products of a particular company. These methods may be usual in the profit-oriented business world, but in a public service profession like medicine they are a new and insidious form of corruption.

Physicians in postgraduate training are also targets, a focus of intensive drug company public relations efforts. A survey reported in *The Journal of the American Medical Association* indicates that physician residents receive an average of six gifts from drug companies each year, gifts intended to fix the name of the drug company in the minds of residents. The gifts may not be expensive, merely pens and notepads, but the gifts do accomplish the purpose of "branding" the drug company in the memory of the recipient.

Another survey reported in the journal *Academic Emergency Medicine* indicates that 41 percent of directors of emergency-department programs allow residents to be taught by representatives of drug companies. Residents receive free drug industry samples at work, and the travel of residents to meetings can be dependent on the availability of drug company support.

Residents, of course, are not in private practice and are not yet full-fledged sales-facilitators for drug companies. Mature physicians are more likely to receive valuable consulting fees and honoraria. A survey by the Henry J. Kaiser Family Foundation found that 92 percent of U.S. physicians received free drug samples from drug companies; 61 percent received free meals, free tickets to entertainment events, or free travel; 12 percent received financial incentives to participate in clinical trials. Of course, most physicians pass free drug samples to their patients, but the free samples are not given for that purpose—the purpose is to create branding and a psychological partnership with the physician.

Financial incentives are also common. Imagine the following scene: An evening in a dining room in an upper-middle-class home, a long dining table, a small dinner party, two attorneys, a politician, maybe a celebrity or two, three or four corporate executives, and a physician with a successful practice. The conversation turns to health and disease, and the physician describes some new treatments and a new drug or two—all in an offhand manner, introducing the names of new drugs to people who will in turn mention the names to many others. The physician has been paid by a drug company to drop the names of these drugs into social settings. The physician may not know much about these drugs, may never have prescribed them for a patient. It doesn't matter. The physician is a sales-facilitator, paid by a drug company to promote its products. Of course the other people at the dinner table know nothing of this arrangement between a drug company and the physician—the other guests are the dupes in this confidence game.

From the standpoint of business relations, the physician is a sales-facilitator, and such facilitators, together with the drug company salespeople who pitch and explain new drugs to physicians, are the shock troops of big pharma in the struggle to extract money from patients. It's the patient who pays the drug company, either directly or through an insurance arrangement.

David Blumenthal, a physician at the Massachusetts General Hospital says,

> Despite the confidence of physicians in their ability to resist efforts by drug companies to affect their behavior—especially in ways that may serve company purposes rather than their own or those of their patients—a substantial body of theoretical and empirical literature . . . suggests that many physicians may be mistaken.

In other words, many physicians find it difficult to serve their patients rather than serve the drug companies that pamper them.

In addition to big pharma's enormous influence over the professional behavior of individual physicians, the pharmaceutical industry pays special attention to medical journals and medical associations, spending an enormous amount of money on journal advertising and medical association promotions, so much so that many journals and medical societies are financially dependent for their existence on the largesse of drug companies and are not likely to do anything that would stop the flow of incoming funds. As the saying goes: Nothing links man to man like the frequent passage from hand to hand of cash.

THE CASE OF THE DRUG NEURONTIN

THE DRUG NEURONTIN PRESENTS AN archetypical case of big pharma manipulation of physicians. Hidden in the back pages of *The New York Times* of December 20, 2002, was the story of Warner-Lambert, a large drug company recently acquired by the larger drug company Pfizer. The news story revealed how the top executives of Warner-Lambert approved a detailed plan by subordinates at its subsidiary Parke-Davis to market the epilepsy drug Neurontin for unapproved ("off-label") uses. Under U.S. federal law, physicians can prescribe drugs for whatever use they believe is in the interest of their patients, but it's illegal for pharmaceutical companies to explicitly market their products for uses that haven't been approved by the U.S. Food and Drug Administration.

Having decided to market Neurontin for non-approved use, the general solution of Warner-Lambert/Parke-Davis to their marketing problem was apparently to use every means available, including financial incentives, to get physicians to prescribe Neurontin for non-epilepsy uses, including prescriptions for migraines, social phobias, and even hiccups. Internal company records from 1995, supplied by a whistle-blower, indicate the company apparently planned to promote Neurontin for these unapproved uses by performing small clinical studies and having the results, only when favorable to the company, published in medical journals. In addition, the company hired doctors to talk to their peers about Neurontin, the physicians expected to present positive messages about the drug, with payment of $500 to $3,000 per speech.

Physicians were offered $200 by the company advertising agency to memorize questions about Neurontin and drop the questions casually into dinner conversations. The marketing program was evidently successful: The sales of Neurontin in 2003 were approximately $3 billion, with more than 80 percent of its prescriptions written for conditions other than epilepsy, with no FDA-approved clinical trials supporting such usage. The U.S. attorney's office in Boston conducted criminal and civil inquiries into the marketing of Neurontin by Warner-Lambert/Parke-Davis. In May 2004, Pfizer pleaded guilty to illegal marketing and agreed to pay $430 million to resolve criminal and civil charges against it.

The Neurontin case came to light only as the result of a whistle-blower, David P. Franklin, a Parke-Davis sales representative. As a whistle-blower, Franklin received $27 million when the case was settled. How many other instances of similar big pharma marketing practices have not been revealed? Every physician paid to promote Neurontin for off-label use was guilty of practicing junk medicine.

Many years ago I was invited to cochair a session at an international pharmacology meeting in Switzerland, travel and lodging prepaid by two large Swiss pharmaceutical companies. Upon arrival at the meeting, I was handed an envelope containing five $100 bills (equivalent today to several thousand dollars), and a note urging me to enjoy my stay in Basel courtesy of the drug companies. As a basic scientist, I had never had such an experience, and I was stupefied. Such was my first contact with the tentacles of big pharma, the octopus that reigns over modern medicine.

BIG PHARMA AND BIOMEDICAL RESEARCH

BIOMEDICAL RESEARCHERS ARE NOT DRUG-COMPANY sales-facilitators, but they are just as important to drug companies because as biochemists and molecular biologists they are the source of new drugs and new patents, because as clinicians they are the professionals who design and carry out the clinical trials that must be completed before a drug can be offered in the marketplace. Many pharmaceutical companies are engaged in pharmacological research with animals, but hardly any drug companies do fundamental research in biochemistry or clinical research with humans.

Most biomedical researchers are employed by universities, and it's usually the researcher's university that has the main relationship with specific pharmaceutical companies. In the past twenty-five years, such relationships have brought both riches and corruption to academia.

One important effect of the Bayh-Dole Act is that many researchers in medical schools and teaching hospitals have an intense focus on possibilities of converting discoveries into financial gain. Suppose you're a biomedical researcher. In the course of your government-sponsored research, have you discovered a new protein that has possibilities as a drug to treat some disease? The first step is to get a joint patent with your university. The second step is to create a small biotechnology company with yourself and your university equity partners. The third step is to license manufacture and distribution of the drug to a large drug company. Everyone makes money. Your university increases its endowment, you become rich, the drug company becomes even richer. Before 1980, results of research sponsored by any branch of the U.S. government were in the public domain. That boring world no longer exists; these days, the most important symbol in biomedical research is the dollar sign. Imagine the pressures that come into play when new data threatens the commercial value of old data.

Questions need to be asked: How often are published data manipulated or fabricated to support commercial interests? How often is scientific data deliberately withheld to protect potential profits? How often are patients endangered by the commercialization of clinical research?

If you want a stark illustration of how things have changed in science, consider this: In the early years of the twentieth century, despite their poverty, and the chance for obvious riches, Pierre and Marie Curie refused to patent their radium isolation process. Concerning the patenting of the process, Marie Curie stated:

> It would be impossible, it would be against the scientific spirit. . . . Physicists should always publish their researches completely. If our discovery has a commercial future, that is a circumstance from which we should not profit. If radium is to be used in the treatment of disease, it is impossible for us to take advantage of that.

These days, the pressure in academia is to translate research into cash. In the year 2000, an article in *The Journal of the American Medical Association* noted that approximately $1.5 billion from industry pours into academic institutions each year. In the life sciences, 28 percent of science faculty members reported funding from private (i.e., nongovernment) sponsors. Several studies have demonstrated an association between single-source sponsorship of clinical research and publication of results favorable to the product of the sponsor.

These funds do enable many kinds of research, but the effect is junk science of vast proportions. Unfavorable results of cancer drugs are less likely to be reported when the study is funded by a pharmaceutical company. Faculty researchers receiving research support from industry are more likely to restrict their communication with colleagues than faculty not receiving industry support, and many faculty who receive gifts from corporate sponsors are subject to prepublication review or data restrictions. Studies have noted a growing number of cases of faculty researchers having personal financial relationships with the companies sponsoring their research. Many such relationships involve paid speaking engagements with payments as high as $20,000 a year, and consulting arrangements that pay as high as $120,000 a year.

The overall situation is not good. A recent review noted that financial relationships among industry, scientific investigators, and academic institutions are pervasive. Lead authors of one of every three articles published hold financial interests related to the article, and approximately two-thirds of academic institutions hold equity in "start-up" businesses that sponsor research performed by their faculty. It's apparent that what our national policies have accomplished is a thorough commercialization of academic biomedical research. One might argue that the money that comes into universities from commercial sources helps defray general expenses. But the small colleges and universities, the places that do need funds but do hardly any biomedical research, don't get much of the pie; the money goes to large research universities with already large endowments.

THE CASE OF NANCY OLIVIERI

THE GENERAL ATTITUDE OF CORPORATIONS is that they are not in business to promote science or advance knowledge, and any scientific results obtained by them internally or through contractual arrangements with scientists outside the corporation are the property of the corporation and can be released to the public only with the assent of the corporation. This attitude is of course in direct conflict with the attitude of academic research that academic science is in general an enterprise in the public interest and that the spirit of science is that of openness between researchers, including rapid sharing of data between laboratories to facilitate scientific progress. Such sharing of data is the norm in most of science. While the corporation wishes to protect its ability to make profits, the biomedical academic world points to a tradition of open communal effort in understanding disease and managing its treatment.

Consider the case of Nancy Olivieri. The thalassemias are a group of chronic inherited anemias characterized by defective hemoglobin synthesis and ineffective red blood cell production. The disease is particularly common in people of Mediterranean descent and in people of African and Southeast Asian ancestry. Thalassemia minor is a relatively benign form of the disease, while thalassemia major is serious and always requires treatment, mainly blood transfusions, and one side effect of these transfusions is an increase of iron in the body to toxic levels.

In 1993, Nancy Olivieri, a physician at the Hospital for Sick Children, a hospital associated with the University of Toronto in Canada, signed contracts with the Canadian pharmaceutical company Apotex to evaluate the use of the drug deferiprone, an iron chelator (a drug that chemically removes iron) that had been used with apparent success with a small number of thalassemia major patients. A problem is that the evaluation of iron chelation therapy in the treatment of this disease is not simple and requires long-term observation of the patient.

The contracts between Olivieri and Apotex included standard confidentiality agreements. A first trial showed promise, and the results were published in *The New England Journal of Medicine* in 1995. More extensive trials were begun in both Toronto and Italy. At that time, Olivieri was the director of the largest hemoglobinopathy clinic in North America.

During the second trials (which apparently did not involve a confidentiality agreement between Olivieri and Apotex), Olivieri noted that extended use of the drug reduced its effectiveness and that the drug had serious side effects — the drug apparently worsened hepatic fibrosis in some patients.

Olivieri sought to report the new findings of adverse events to the hospital review board, present them at a scientific meeting, and submit them for publication. In response, Apotex stopped all clinical trials that involved Olivieri and threatened to take legal action if Olivieri released to the public the information from the second trial. Eventually, lawsuits totaling $20 million were lodged against Olivieri. At the time all of this happened, the University of Toronto was evidently negotiating a corporate donation from Apotex of $12.7 million for a new medical building. The university declined to provide legal support for Olivieri, who ignored threats of legal action against her and published her findings in *The New England Journal of Medicine* in 1998. In 1999, Olivieri was dismissed as director of the hemoglobinopathy program at the University of Toronto's Hospital for Sick Children. A public scandal erupted and an inquiry board was formed to investigate the case. Olivieri was finally vindicated, exonerated in 2001 of misconduct, and rehired by the University of Toronto.

This case of conflict of principle between an academic researcher and a pharmaceutical company is unusual only because it received a great deal of publicity and one of its main issues — patient safety — was quite clear. When two entities whose missions are intrinsically different seek to derive mutual benefit from a relationship, the stage is set for potential conflict. But considering the financial resources available to drug companies, conflict, when it happens, is never conflict between equals. The University of Toronto ought to have supported Olivieri on principle. Apparently, principle had been corrupted by academic association with corporate interests, and since the University of Toronto is a major institution with a most laudable record of achievement in biomedical research, one must wonder how many unpublicized cases similar to this one have bedeviled academic science during the past twenty-five years. An editorial in *The New England Journal of Medicine* concluded:

> *The Olivieri case represents an important warning that academic freedom can disappear if we do not protect it. How tempting and comforting it would be to believe that the case is unique. And how much wiser it would be to conclude instead that there, but for the grace of God, go we.*

The fundamental question today is whether biomedical science is in the process of being hopelessly compromised by its growing dependence on financial support from the pharmaceutical industry. At the present time, an estimated 60 percent of research funding in the biomedical sciences comes from private sources. One might argue that in the seventeenth and eighteenth centuries rich patrons were often the sole source of support for many scientists, and those centuries produced brilliant ground-breaking science that endured and remained devoted to the search for new knowledge. The difference is that for the most part those rich patrons themselves were interested in new knowledge and not in profits from science. If some good science results from dependence on commercial interests, how much science is twisted by such dependence?

THE DANGERS OF COMMERCE IN SCIENCE

OUR TIME IS OUR TIME. In the past, the arena of financial frenzy was land, then transportation, then manufacturing. These days the arena of financial frenzy is science—computer science and biological science and materials science. With a suddenness that has left many older scientists dizzy, science has moved to center stage as a focus of commerce. Money flows. A river of money flows into the fields of the knowledge-producers, promotes growth in some places and greed in other places. Stand back and consider history and maybe it's all predictable. In 1840, Alexis de Tocqueville wrote of the impact of American democracy on American science: "As soon as the multitude begins to take an interest in the labors of the mind, it finds out that to excel in some of them is a powerful method of acquiring fame, power, or wealth."

The Bayh-Dole Act has changed American science, pushed science and universities in a direction not in the public interest. During the past twenty-five years, American academic science and medicine have become more closely tied to commercial interests than ever before. Biologists began scrambling to exploit new applications of genetic engineering and were recruited to serve on scientific advisory boards of new companies built with venture capital and the promise of new drugs and therapies. Ten years after passage of the Bayh-Dole Act it was estimated that 1,000 university-industry research centers had been established at over two hundred American universities, more than doubling university-industry partnerships in one decade. Many scientists partnered with business school graduates, started their own companies while retaining their academic appointments. The result of all of this has been a sea-change transformation of biomedical science into an enterprise governed by market forces. The loss of independence of academic research can never be good for the public, will always foster corruption of science, and is already doing substantial damage to society.

This is our era and we cannot go backward. The age of bedside doctoring in the middle of the night is past, the image a museum-piece of medical history. The age of the Curies is also gone, a memory of an exemplary devotion to science. We live in a time when scientists strive for large financial rewards for their work, and it's doubtful that will change in the near future. American science has become a profit-oriented endeavor. The problem, however, is not so much profit itself but what the striving for profit by both industry and universities does to science. We need to struggle to ensure that the quest for financial gain does not corrupt the scientific enterprise to a point of destruction. So far, the outcome of our struggle does not look promising.

8

QUACK DOCTORING:
IT'S A FREE COUNTRY, ISN'T IT?

Hell, if I hadda knowed a feller had to git up every night I
would never have started to learn doctoring.
— AMERICAN MEDICAL SCHOOL GRADUATE, CIRCA 1875

Imagine a survivor of a failed civilization with only a tattered
book on aromatherapy for guidance in arresting a cholera epidemic. Yet, such a book would more likely be found amid the
debris than a comprehensible medical text.
— JAMES LOVELOCK

MOST PEOPLE CONSIDER THEIR personal health the most important
element of their lives, and the way people of a nation deal with their
personal health can reveal something about their general attitudes. If
there is any doubt that we are *not*, as some claim, in an "age of science," consider the fact that 60 percent of Americans use nonscientific medicine ("alternative" medicine) either exclusively or together with standard evidence-based
scientific medicine.

The U.S. government spends an enormous amount of time, effort, and
money to train science-based medical personnel, support science-based medical facilities, and develop science-based treatments, and yet the same government, in 1994, deregulated unscientific quack medicine, not to improve public
health, but apparently to make the public more vulnerable to medical "free enterprise."

Anyone keeping track of the arduous climb of science-based medicine to its
present understanding of health and disease can wonder with reason if the
people who make the laws have all gone mad.

QUACK MEDICINE IN AMERICA

IN 1875, THERE WERE 460 "medical schools" in the United States (nearly four times as many as now), most of them diploma mills whose main function was to collect tuition fees. Students took courses consisting of two four-month or six-month terms at approximately $60 a term, and often the second term was a verbatim repetition of the first term. These schools were staffed largely by "professors" who had graduated from similar institutions. A few university medical schools were of a higher caliber, but not much higher. In 1869, the dean of Harvard Medical School explained that the medical school had no written examinations because "a majority of the students cannot write well enough."

If nineteenth-century American medicine seems like medicine in another country, current quack medicine in America is a raucous bazaar on another planet. Acupressure, acupuncture, aromatherapy, Ayurvedic medicine, biofeedback, chelation therapy, chiropractic, Christian Science, herbal medicine, homeopathy, hydrotherapy, iridology, macrobiotics, massage therapy, meditation, megavitamin therapy, moxibustion, naturopathy, osteopathy, relaxation techniques, rolfing, shiatsu, tai chi, yoga—these are all types of what is called "alternative medicine," an extensive repertoire of techniques whose efficacy is largely unproven and whose methods are often dangerous—either by direct effect of the technique or as a result of a delay in standard treatment.

The name "alternative medicine" is itself a misrepresentation: There is no such thing as "alternative" medicine, there is only evidence-based scientific medicine standing apart in a large courtyard filled for the most part with quacking entrepreneurs hawking various versions of junk science, snake oil, and hokum.

If you think alternative medicine is merely a minor preoccupation of the American public, the reality is that the total annual cost of alternative medicine in the United States exceeds $27 billion, comparable to the total cost of standard medicine. In 1998, a survey estimated that 630 million visits were made to alternative medicine practitioners, which exceeded the total number of visits that year to all American primary care physicians.

AYURVEDIC MEDICINE

FIVE-THOUSAND-YEAR-OLD AYURVEDIC MEDICINE IS CURRENTLY a growing phenomenon in the United States. Approximately 800 million people use

Ayurvedic medicine in India and uncounted thousands in the Western world. In the United States, Ayurvedic remedies are now available from South Asian food markets, Ayurvedic practitioners, health-food stores, and the Internet. Because Ayurvedics are marketed as dietary supplements, they are "regulated" under the Dietary Supplement Health and Education Act, which does not require proof of safety or efficacy.

Ayurvedic medicine sets the paradigm for nearly all the other quack regimens in common use in America. It's a medical approach rather than a single medicinal, an approach using herbs, minerals, and metals—and it's dangerous. In particular, lead toxicity has been associated with Ayurvedic medicine, with consequences including epileptic seizures, fatal infant brain damage, congenital paralysis, sensori-neural deafness, and delay in childhood development. Since 1978, at least fifty-five cases of heavy metal intoxication associated with Ayurvedic herbal medicine products in adults and children have been reported in the United States and abroad, with an unknown number unreported.

A recent study in *The Journal of the American Medical Association* concludes that one-fifth of Ayurvedic herbal medicine products produced in South Asia and available in South Asian grocery stores in the Boston area contain potentially harmful levels of lead, mercury, or arsenic, and that users of Ayurvedic medicine may be at risk for heavy metal toxicity. Lack of regulation of such herbal medicinals may benefit the commercial entities that import and market these products, but society pays heavily for the consequent economic costs of health care for poisoned children and adults. As usual, children are the most vulnerable, with lead and mercury poisoning, even at minuscule concentrations, producing debilitating effects on the developing central nervous system.

THE DIETARY SUPPLEMENT HEALTH AND EDUCATION ACT

THE DIETARY SUPPLEMENT HEALTH AND Education Act (DSHE Act) passed in 1994 restricted the Food and Drug Administration's control over dietary supplements, leading to enormous growth in their promotion. Between 1990 and 1997, use of herbal remedies increased 380 percent in the United States. A 1998–1999 survey estimated that 14 percent of U.S. adults took herbal supplements and approximately one in five adults taking prescription medications also used an herbal or dietary supplement. By the year 2003, the dietary supplement industry had assumed a substantial proportion of the health market, grossing nearly $18 billion in 2001.

The issue here is one of definition. The DSHE Act defined "dietary supplements" as vitamins, herbs, minerals, amino acids, and any article said to "supplement" the diet. Before the act, such substances were considered foods or food additives, and claims in advertising that they improve health could be based only on the nutritive value of the products. The act permits health claims in advertising based on "structure and function" rather than nutritive value, and as a result these products are no longer considered food additives, and no premarketing screening for product safety, effectiveness, and quality is required of the manufacturer.

Manufacturers of these products can claim a product enhances normal "structure and function" of the body or any particular organ or part of the body, without submitting supporting evidence to the Food and Drug Administration or to any other regulatory agency. For example, one can legally claim that a product promotes "cholesterol health," but it is illegal to claim without evidence that the product reduces elevated cholesterol levels or prevents myocardial infarction. As one might expect, the marketing of these products then becomes a game of manipulation of semantics and clever use of words.

Since passage of the DSHE Act, sellers of these products use direct consumer advertising to promote health claims, and the products are purchased primarily in "health food" stores, grocery stores, and by mail order on the Internet—all without medical supervision.

THE PUBLIC, OF COURSE, is confused about protections. The public is protected against adulteration of ordinary foods, but not protected against adulteration of "health" food supplements sold in a health-food store next door to the supermarket. The public is protected against high toxicity of standard drugs, which are immediately withdrawn as soon as such toxicity is discovered, but not protected against high toxicity in herbals, many of which contain toxic concentrations of heavy metals whose effects are insidious and long-term. The public is protected against false claims of efficacy of standard drugs, but not protected against false claims of efficacy of "alternative" drugs, as long as such drugs do not refer to a specific disease by name. But such specific referrals to diseases are the rule rather than the exception in alternative-medicine advertising, so evidently the public is not protected against false claims of any substantive kind.

Too many people believe that if an herbal product sold in a health-food store were poisonous, "they wouldn't let it be sold." But who is "they"? The Food and Drug Administration? The health-food shop owner? The distributor of the product? The manufacturer in Brazil or India? Most certainly, for liability reasons, a product would be withdrawn if people ingested the product inside a health-food store and then immediately dropped dead on the sidewalk outside the shop. But if the killing process is slow, the operative idea is apparently "caveat emptor"—buyer beware. In short, you are not protected.

HERBAL MEDICINES

THE VARIETY OF HERBAL MEDICATIONS available to the public is enormous. The most important of these is probably ginkgo (also called ginkgo biloba). It's the herbal medication first in sales in health-food stores, and it has a 2,000-year-old history in Chinese medicine. The product is derived from dried leaves of the Ginkgo biloba tree, which was originally native to China, but which is now cultivated wherever possible because of its commercial value. Advertising for ginkgo claims: "Its effects in improving circulation also contribute to its use for impotency and peripheral vascular insufficiency. Ginkgo treats depression, headaches, memory loss and ringing in the ears (tinnitus). It is also recommended for Alzheimer's, asthma, eczema, heart and kidney disorders."

The trouble with ginkgo is that none of the claims for it have been verified by extensive controlled clinical studies, and the ginkgo herbals sold to consumers are not uniform, not regulated in the United States, and often of uncertain origin. Ginkgo has several ingredients thought to be pharmacologically active, including flavonoid glycosides that are believed to have some antioxidant activity and to reduce capillary fragility, and a terpene called ginkgolide B, which may affect platelet aggregation. Although a few minor studies have shown apparent modest effects on cognitive function in the elderly, there is no scientific basis for the broad claims that ginkgo is effective in heart and kidney disorders.

Another major herbal is St. John's wort. The name "St. John's wort" is given to a variety of plants of the genus *Hypericum*, the plants bearing yellow flowers and called St. John's wort after the medieval custom of gathering them on St. John's Eve to ward off evil spirits. The active ingredients include napthodi-anthrones, flavonoids, and xanthones, and the highest potency is supposedly found in the flowering tops of the plants. The claims for this herbal are as impressive as for other herbals: "St John's wort is effective in the treatment of mild to moderate depression. Recent studies have shown that it could have a potent antiviral effect against enveloped viruses."

Many facts about St. John's wort are not revealed to the public in commercial promotions. For example, the active ingredient hypericin absorbs light in the ultraviolet and visible ranges and is phototoxic to skin. A controlled study with human lens epithelial cells has demonstrated that cells exposed to hypericin in combination with ultraviolet-A or visible light undergo necrosis and apoptosis—the cells are killed. The researchers suggested that St. John's wort is potentially phototoxic to the eye and could contribute to the genesis of early cataract.

Another problem with St. John's wort concerns interactions with standard drugs. For example, St. John's wort can cause substantial decreases in plasma concentrations of a range of co-administered drugs, influencing the effectiveness and safety of drug therapy. St. John's wort is not a weakly acting herbal. It's touted as an antidepressant as powerful as Prozac (fluoxetine). A physician's prescription is required for Prozac in any American pharmacy, but it's not required for St. John's wort in the health-food shop next door. Are prescription requirements for standard drugs a regulatory device to protect patients against overuse of potentially dangerous drugs, or merely a device to provide certain drugs with a competitive edge against herbals? If St. John's wort is truly as powerful as Prozac as a neurotropic drug, why isn't St. John's wort regulated?

In general, interactions between herbals and standard medications are a serious problem for the public. Herbal medicines are almost always mixtures of more than one active ingredient, a multitude of pharmacologically active compounds, and this increases the likelihood of interactions with standard drugs. The likelihood of interaction is greater than that between standard drugs themselves, since standard drugs usually contain only one active chemical entity. The medical community is becoming aware that the use of herbal agents by patients may represent a potential risk to patients undergoing conventional pharmacotherapy.

The product "echinacea" is another herbal remedy. It's widely used and has more than $300 million in annual sales. The plants Echinacea pallida, E. augustifolia, and E. purpurea are native to North America and are also known as "purple cone-flowers." Active ingredients include isobutylamides, chicoric acid, polyenes, alkaloids, and alkylamides. Echinacea claims are as broad as claims for other herbal medicines:

Because it has natural antibiotic actions, echinacea is considered an excellent herb for infections of all kinds. In addition, it works to boost lymphatic cleansing of the blood, enhances the immune system, and has cortisone-like properties which contribute to its anti-inflammatory action. It's recommended for stubborn viral infections, yeast infections, and for arthritic conditions.

None of these claims are scientifically substantiated. In July 2005, a controlled study of 437 volunteers reported that extracts of echinacea root, "either alone or in combination, do not have clinically significant effects on infection with a rhinovirus [common cold virus] or on the clinical illness that results from it."

The list of other herbal miracle medicines is almost endless. A combination of ginseng, garlic, and Tien-Chi is touted as "potentially beneficial for AIDS, radiotherapy patients, and chemotherapy patients, as it reduces the side effects of toxic drugs by increasing red and white blood cell counts." Dang Shen is "given for breast cancer, asthma, diabetes, heart palpitations, memory or appetite loss, and insomnia." Saw palmetto is recommended as a "preventive and therapeutic agent for benign prostatic hyperplasia." Kava kava is offered as a "valuable urinary antiseptic, helping to counter urinary infections and to settle an irritable bladder. Kava kava's analgesic and cleansing diuretic effect often makes it beneficial for treating rheumatic and arthritic problems such as gout." Valerian root is claimed to be "the most effective in treating a wide range of stress conditions such as irritability, depression, fear, anxiety, nervous exhaustion, hysteria, delusions, and nervous tension. . . . The herb is useful for treating shingles, sciatica, neuralgia, multiple sclerosis, and epilepsy."

In short, name your disease and the clerk in the health-food shop will pull out an array of herbals for "treatment." Buy this one or that one. Buy four different herbals and get a discount. And you don't need to pay a physician to get a prescription.

It's important to understand that the perception that botanical products used as folk remedies are inherently safe simply because they are plants is based more on wishful thinking than on science. Evidence of the toxicity of such products has accumulated over the years, which is not surprising, since botanicals are complex mixtures of chemicals, and many of these chemicals are potentially toxic. The Food and Drug Administration was recently compelled to issue warnings about nephrotoxic, hepatotoxic, and carcinogenic effects associated with botanical products containing kava, comfrey, and aristolochic acid— all herbal remedies used widely in the United States and Europe.

A serious problem is that many herbal products contain undisclosed prescription or over-the-counter drugs mixed with heavy metals. In 1998, the California Department of Health reported that 32 percent of Asian patent medicines sold in that state contained undeclared pharmaceuticals or heavy metals. The drugs most frequently found were ephedrine, chlorpheniramine, methyltestosterone, and phenacetin; 10 to 15 percent contained lead, mercury, or arsenic. In another study, more than five hundred Chinese patent medicines were screened for the presence of heavy metals and 134 drugs. Approximately 10 percent were found to contain undeclared drugs or toxic levels of metals.

We have struggled through thousands of years of medical history, acquiring scraps of medical knowledge piece by precious piece, and to have this struggle and its fruits tossed aside for the sake of a so-called free market economy is an intellectual travesty that ultimately causes hardship, pain, and death.

Is it possible that a solution to the problems of American health care is to throw away modern medicines and just return to digging herbs out of the ground? If we couple that with the ancient practice of drilling holes in the skull to cure mental illness we might be able to reduce our taxes still further.

As always, junk science bamboozles the public. Allowing these products to be sold unregulated as "food supplements," rather than having them recognized and regulated as medicines, makes a mockery of Congress as it postures as a protector of the American public.

WHO USES ALTERNATIVE MEDICINE?

WHO USES ALTERNATIVE MEDICINE? SOME sociologists have reported that users of alternative medicine are primarily members of uneducated minority populations, particularly Hispanic and South Asian. In contrast, a study reported in 1998 found that users of alternative medicine tend to be better educated and to hold a "holistic" orientation toward health (an orientation that stresses the fused importance of body, mind, and spirit). The study reports that these people are more likely to have had some type of "transformational experience" that changed their worldview. The study also found that such users of alternative medicine are in poorer health than nonusers. In general, assessing the sociological profiles of users of alternative medicine is made difficult by the variegated nature of alternative medicine, with some forms of alternative medicine appealing to specific minority ethnic groups, while other forms of alternative medicine appeal to the large numbers of middle-class people with better education who constitute the "New Age" group that focuses on meditation, various types of massage therapies, and exotic Asian health regimes. So-called better education obviously doesn't preclude being fooled by junk science—which emphasizes why government regulation of alternative medicine is needed.

But of course the advocates of alternative medicine oppose regulation. Some advocates have argued that many alternative therapies cannot be subjected to the standard scientific method of testing medical treatments and therefore must rely on anecdotes, beliefs, theories, testimonials, and opinions to support effectiveness and justify continued use. This is a specious argument that can appeal only to individuals ignorant of science or befuddled by anti-science attitudes. Any medication or medical treatment is either effective or in-effective, and a controlled experiment can easily be devised for anything offered to the public. The argument against testing is maybe a convenience for entrepreneurs selling alternative medicine products and treatment regimens, but the argument does a great disservice to the public and the potential for serious public danger remains enormous. Using the same argument, one could advocate a return to tribal witch-doctoring as a method of curing disease.

THE CASE FOR REGULATION

WHAT IS THE RATIONALE FOR the current system? Why regulate standard science-based medicine and drug treatment and not regulate so-called alternative medicine? If alternative medicine deserves to be considered parallel to standard medicine, then what reason could there be not to regulate it? And if alternative medicine is merely quack medicine, then certainly, as potentially harmful quack medicine, it should be *more* regulated and not less regulated than standard medicine. Or, perish the thought, let us deregulate all of medicine and pharmaceuticals and return to the days of horse and buggy "doctoring" by diploma-mill doctors. At least we would then have the satisfaction of consistent attitudes.

Apart from the risk of fraud and misrepresentation perpetrated on the public, the substantive dangers of alternative medicine warrant regulation. The lack of standardization of alternative medicine products, particularly herbal medicines, is extremely dangerous. The active ingredients of many of these products vary from one pill to the next in the same package, and between suppliers the situation is even worse. Chemical contents, whether toxic or benign, are not regulated in these products, and the consumer is at the mercy of the product.

Another important danger is produced by adulteration, since no law prevents these products from mixture with anything the manufacturer fancies. There is evidence that in some cases Viagra (sildenafil) has been added to herbals. One needs a prescription from a standard physician to obtain Viagra, but the same Viagra can be obtained without a prescription in an adulterated herbal "medicine" whose label claims it contains only "natural" ingredients. A similar situation exists with regard to the prescription drugs glyburide, colchicine, adrenal steroids, alprazolam, phenylbutazone, and fenfluramine (which has been withdrawn as a standard prescription drug after findings of severe cardiac damage in some patients).

As previously noted, a most serious problem, and a problem that's often hidden, is potential interaction between herbal medicines and standard drugs in patients who use both. Patients who are using herbals often do not report this to their physicians, who may prescribe standard drugs that in combination with herbals can do great harm. And even if a patient does report the use of herbals to a physician, most physicians are unfamiliar with herbal medicines, particularly since relatively few studies of herbal medicines are published in widely read medical journals.

And from the standpoint of general public health, that's the most outstanding problem: In contrast to the way the medical community monitors the negative effects of standard drugs, there is very little reporting in journals of adverse events with herbal medicines and alternative medicine treatments. For the most part, the public today is as much at the mercy of snake oil salesmen as it was in the nineteenth century.

Regulation of food and drugs has always been strongly resisted by industry, and Congress has usually acted only in response to strong pressure from the public. The Food and Drug Acts passed in the twentieth century, which provided important protection to the public, were subverted by the passage of the DSHE Act in 1994. Indeed, a recent assessment in *The New England Journal of Medicine* states:

This misguided legislation freed the dietary supplement industry from effec-
tive oversight by the FDA, transferring the burden of proof for establishing
the safety of herbal medicines from the manufacturer to the FDA. Dietary
supplements are now subject to lower safety standards than food additives.
Consumers are provided with more information about the composition and
nutritional value of a loaf of bread than about the ingredients and potential
hazards of botanical medicines.

CHIROPRACTIC AND CHIROPRACTORS

AN ARGUMENT CAN BE MADE that any "alternative" medical treatment is potentially dangerous if used as a substitute for evidence-based standard treatment.
In general, this is a valid point, which means that it's not the apparent lack of
harm of a particular treatment that needs to be the focus, it's the question of
whether, for some particular patient, an apparently harmless "alternative" treatment delays standard treatment for a progressive dangerous medical condition.
Patients who have been told there is no standard treatment for their condition
and who seek out alternative medicine as last-resort medicine present one issue,
an issue of commercial exploitation. In contrast, people who seek out alternative medicine as a complete substitute for standard medicine present an entirely different issue, an issue of public danger.

The alternative medicine now called "chiropractic" was founded in 1895 by
Daniel David Palmer, a quack without medical training who ran a grocery store
in New Boston, Illinois, before moving to Davenport, Iowa. In Davenport,
Palmer hung out a shingle as a healer, and the story is that he used spinal manipulation to instantly cure his first patient of deafness.

Today, an estimated 50,000 practitioners of chiropractic have offices in the
United States, with approximately 14,000 students in training at sixteen chiropractic colleges. It's estimated that by the year 2010, the number of chiropractors in the United States will exceed 100,000. There are now more than 10,000
chiropractors in the state of California alone.

Two main modes of chiropractic exist, with substantial differences that
cause public confusion and danger. The classical "orthodox" mode of chiropractic essentially maintains that all diseases are diseases of the spine and joints.
In contrast a "reformed" chiropractic prefers a tangential contact with scientific

medicine, maintaining that the classical notion defining all diseases in terms of spinal and joint dysfunctions is obsolete, and that chiropractic spinal and joint manipulations may be helpful in treating only certain conditions. The two modes of chiropractic are in sharp conflict. The National Association of Chiropractic Medicine will accept only members who *renounce* classical chiropractic. So it behooves anyone who visits a chiropractor's office to inquire where in this muddied battlefield the practitioner at hand is positioned.

In theory, "chiropractic" is a therapeutic system that uses the recuperative powers of the body and the relationship between the musculoskeletal structures and functions of the body, especially as they involve the spinal column and the nervous system, in the restoration and maintenance of health. The most common method of chiropractic manipulation is a short lever, high-velocity thrust directed specifically at a "manipulable lesion." This procedure is typically performed with the patient lying on his or her side on a segmental table.

In addition to the many individuals in the United States who are licensed and certified to practice chiropractic, there is also extant in the United States the "McKenzie method of physical therapy," in which patients are placed in one of three broad categories ("derangement," "dysfunction," and "postural" syndromes) that determine therapy. In the McKenzie formulation, most back, buttock, and leg pain is considered to result from a derangement syndrome treated by exercises that "centralize" pain from the feet, legs, or buttocks to the lower back. Patients are taught to perform exercises that centralize their symptoms and to avoid movements that peripheralize them. The McKenzie method relies on patient-generated forces and emphasizes self-care.

Does it work? Do either chiropractic or McKenzie manipulations have a scientific basis? Can these alternative treatments cause a delay in treatment of potentially dangerous conditions? The first question is easier to answer. In 1998, an article in *The New England Journal of Medicine* reported a study involving random assignment of 321 adults with persistent low back pain. Subjects were assigned to a McKenzie group, to a chiropractic group, and to a minimal intervention group that received only an educational booklet. All McKenzie physical therapy and chiropractic patients were treated by certified practitioners of those methods for one month with as many as eight visits per patient during that time interval. The study received the cooperation and assistance of local McKenzie therapy and chiropractic professional groups.

The results of this study were that for patients with low back pain the McKenzie method of physical therapy and chiropractic manipulation had similar effects and costs, and that patients receiving these treatments had only marginally better outcomes than those patients receiving the minimal intervention of an educational booklet.

The average costs of treatment (in Seattle) were $238 per patient for the McKenzie group, $226 per patient for the chiropractic group, and $1 per patient for the booklet group.

Health care is business. Quack medicine is big business. As usual, the public bleeds with both its money and its health.

NATUROPATHY

THE PRACTITIONERS OF NATUROPATHY CALL themselves naturopathic "physicians" and say they practice naturopathic medicine. There are currently in the United States three schools that train naturopathic physicians: one in Seattle, Washington; one in Portland, Oregon; and one in Tempe, Arizona. All three colleges offer a four-year program leading to a "Doctor of Naturopathic Medicine" (N.D.) degree. The first year of the training program focuses on basic sciences. The second year focuses primarily on pathology and laboratory diagnostics. The third and fourth years focus on various categories of standard clinical medicine, and also on "naturopathic therapies" such as botanical medicine, homeopathy, counseling, therapeutic nutrition, and physical medicine. After all of this, students can choose advanced elective courses in subjects such as acupuncture or Ayurvedic medicine. As of 1998, naturopathy practitioners were licensed in only eleven states.

Apart from the course material in alternative medicine, the training program in naturopathic medicine has the appearance of a training program in standard medicine: basic biomedical sciences followed by training in clinical medicine. But it's only an appearance: A student in a naturopathic training program never gets near a standard hospital patient with any sort of standard disease, never witnesses or participates in any serious invasive procedure (except perhaps acupuncture), never learns to write a prescription for standard drugs because the student, once a naturopathic "physician," will never be allowed by law to write such a prescription. If this is medicine, it's bloodless and drugless medicine, although it can be argued that use of the term "medicine" may not be appropriate. The basic approach of naturopathy is apparently to treat the "whole patient" with available "natural" remedies, which in practice seems to mean any remedies not forbidden to the naturopathic practitioner by law, which in turn means not any remedy whose use requires a standard physician's training and license. The naturopath wears a white coat, has a nurse who wears white, has framed certificates on the wall, all in an office carefully designed to look "clinical."

One of the major texts in naturopathy is the Murray-Pizzorno 950-page volume, *Encyclopedia of Natural Medicine*. In fine print behind the title page, one finds the disclaimer, "The information presented herein is in no way intended as a substitute for medical counseling." The disclaimer is advisable, since it's against the law to provide medical counseling without a license to practice such counseling. Nevertheless, the title of this book, sold primarily to consumers, presents the book as an encyclopedia of "natural" medicine. Not naturopathy, but "natural" medicine. (What is "unnatural medicine"?)

The publishers state they have sold 400,000 copies of this book, which disclaims its use as a substitute for medical counseling, but which by title and contents appears to present itself to consumers as indeed medical counseling, with explicit naturopathic treatment advice on standard diseases such as AIDS, Alzheimer's disease, anemia, angina, asthma, manic depression, cystitis, bronchitis, pneumonia, celiac disease, cervical dysplasia, depression, diabetes, fibromyalgia, glaucoma, heart disease, hepatitis, hypothyroidism, inflammatory bowel disease, macular degeneration, multiple sclerosis, osteoporosis, periodontal disease, psoriasis, rheumatoid arthritis, ulcers, and so on.

Few of these naturopathic recommendations have been adequately tested or proven beneficial according to current biomedical standards. Among recommended treatments for the eye disease glaucoma, for example, are not any standard drugs currently used in standard medicine to lower intraocular pressure (the naturopath cannot write prescriptions for such drugs), but two botanical "medicines," bilberry extract three times per day, and ginkgo biloba twice daily. The research literature reference in the text for ginkgo biloba treatment of glaucoma is a single twenty-five-year-old paper published in an obscure German journal. (Never mind, the man in the white coat will sell you a bottle of ginkgo biloba at a low price and you don't need a prescription for a refill.)

If the public is confused about naturopathy and "natural" medicine, and confused about practitioners with "N.D." rather than "M.D." after their names, their confusion is warranted. Meanwhile, government offices provide little help or protection, and state and federal regulatory agencies seem as confused about their function as the public is confused about health and medicine.

❀ ❀ ❀

THE MOST IMPORTANT PROBLEM concerning alternative medicine in America is that blatant junk science is allowed to prosper, not for the good of the public but for the good of special business interests with strong lobbies in Congress. Public interest is sacrificed for private interest, and the result is a wild medical free-for-all that puts too many people in the hands of charlatans. There's no logic to it. Does it make sense to closely regulate standard medicine and have hardly any regulation at all for nonstandard medicine? If anything, the regulations for alternative medicine should be the most stringent possible, limiting claims and practices in order to protect the public from the worst kind of junk medicine — quack doctoring.

Maybe sick people living in 1875 had an advantage of clarity about medicine: At least they could assume that no kind of doctor knew anything of any importance.

9

HEALTH CARE IN AMERICA:
SORRY, NOT FOR EVERYONE

America's medical professionals are the best and brightest in
the world, and set the standard for the world.
—PRESIDENT GEORGE W. BUSH, THE WHITE HOUSE, 2005

The Institute of Medicine estimates that between 44,000 and
98,000 Americans die each year from medical errors. Many
more die or have permanent disability because of inappropriate treatments, mistreatments, or missed treatments in ambu-
latory settings.
—PRESIDENT GEORGE W. BUSH, THE WHITE HOUSE, 2005

HEALTH CARE IS APPLIED science, and when health care in a society is
twisted by politics, prejudice, or profiteering, the result is junk science
destructive to the public. You may think the health-care issue is solely a
political issue or a social issue, but applied science is science, and indeed the
two can't be separated. The issue of health care is frequently discussed in biomedical journals because biomedical scientists recognize that it's an applied science issue. When facts about health care (or the lack of it) are twisted to serve an
agenda, the twisting produces junk science no less than twisting facts about the
safety of explosives produces junk science. I'm not concerned here about the social or political aspects of health care (although some of my biases may be evident). My concern is the corruption of a major applied science, and thereby the
corruption of reason in an issue of maximum concern to the American public.

We've had the important questions in front of us for nearly fifty years. What
is the status of public health in America? Is the current health insurance structure adequate for the needs of the people? Is it myth or reality that poverty produces bad health? Are all people equal in health care, or does racial and ethnic

discrimination contaminate the system? Is the current American system of health care the best in the world? Is it myth or reality that more people in the United States are killed each year by medical errors than by automobile accidents? Would universal health insurance solve the problems of access to health care? Is it really possible to have a completely free-market health care system in the United States?

Sure, every one of these questions has its political and social components, and right down to the heart of things, since in America the health-care issue is about as political as you can get. But what I want to do here is get away from the politics and focus on health-care facts that are usually buried in the heat of wrangling about it. I want to focus on the questions that need answers.

FIRST OF ALL, THERE'S a premise that should be made explicit. My assumption is that finding a way to apply science to care for the health of as many people as possible is a good thing to do. As an idea, this is only an extension of what people in a family try to do for themselves — look out for each other. My premise is that as a nation our goal is to do the same for the people at large — in this case by the application of medical science.

It's a high-minded objective with grueling difficulties. One immediate and central problem is that in a free-market health-care system such as that current in the United States, a substantial fraction of the population cannot afford the health care that they urgently need. No money, no health care. If you barely have money to put food on the table, you don't see a doctor or dentist unless it's an absolute emergency. And that's not all. There's plenty of evidence that poverty also makes people sick and often makes them even poorer because they can't work. My view is that at the present time the issue of poverty is inseparable from the issue of health care.

But here I'm looking at public health as a problem in medical engineering, and my focus is how this particular applied science gets twisted in the public arena.

PUBLIC HEALTH IN AMERICA

THE CHARACTER OF PUBLIC HEALTH in America is not what most people think and not what is usually described to them by politicians and the media. Statistics are manipulated, data is concealed by one government agency or another, the media does its usual job of emphasizing what it thinks will keep the readers and viewers coming back for more. The consequence is a mythology about American health care that pervades every class and every locality.

For example, people generally believe that life expectancy in America is high for everyone regardless of where they live, but the data indicate otherwise. People born in certain rural counties of Minnesota, Colorado, Iowa, or Wisconsin will live on average twenty-five years longer than people born in four counties of South Dakota, twenty-three years longer than in twelve counties in Mississippi and Alabama, and twenty-two years longer than people born in Washington, D.C., or Baltimore.

The variance in life expectancy in the United States between women of Japanese extraction in Bergen County, New Jersey, and women of Japanese extraction in Bennett County, South Dakota, is forty-one years, an enormous difference and one that defies understanding in terms of racial, ethnic, or economic differences.

In general, overall disparities in life expectancy between different parts of the United States are greater than for any other nation in the world, and the reasons for this are not clear.

HEALTH INSURANCE

APART FROM THE CHARACTER OF public health, the problems of health insurance continue to fester in American society. The lack of health insurance by nearly 18 percent of the American population is a serious structural problem that needs attention. The insurance gap affects working families (those with at least one worker) particularly hard. A full 80 percent of uninsured people under age sixty-five live in working families, many surviving on low wages paid by service jobs. Families with two wage earners do better, but 10 percent of those families lack health insurance. One can take it as a given that any working-class family without health insurance sharply limits visits to physicians and clinics for preventive health care or for potentially serious illness, visits that families that do have health insurance consider ordinary. What is the acquiescence of society to the existence of this insurance gap except a signal that the economically comfortable class believes 40 million Americans are expendable?

It's clear, and has been clear for decades, that people without access to preventive health care and without access to adequate treatment of disease once disease appears, are less likely to lead productive lives, are more likely to die early, and are certainly more likely to be a major financial drain on the national economy.

HEALTH AND POVERTY

THE HEALTH CONSEQUENCES OF POVERTY are complex and involve more than lack of access to health care. Poverty itself debilitates health, makes health care necessary, and the lack of adequate health care soon increases debilitation, produces more serious disease, and ultimately magnifies the cost to society. The game of conservative politicians is apparently one of sacrificing a large segment of the American public in order to completely preserve free-market medicine, a sacrifice never explicitly identified by these politicians to the public.

Of the rich countries of the world, the United States has the largest gap between rich and poor. In contrast to the attention paid by American politicians and the American media to racial differences and racial inequalities, class differences and class inequalities are largely ignored. Many people view class disparities within racial and ethnic groups as an inevitable consequence of capitalism, or a consequence of "inappropriate" culture, or simply a matter of bad luck. But no matter how class differences are viewed, class differences exist, and differences in rates of premature death, illness, and disability are closely related to socioeconomic status. In general, people in lower classes die earlier than people in higher socioeconomic classes, and the pattern persists from the poorest to the richest classes. In the time frame 1972 to 1989, people who earned less than $15,000 a year (in 2003 dollars) were three times as likely to die prematurely as were people earning more than $70,000 a year.

Rural America is particularly vulnerable to the deficiencies in the healthcare system. In 1999, 14 percent of rural Americans lived in poverty, compared to 11 percent of urban Americans. Rural communities have fewer hospital beds, fewer physicians and nurses, and fewer medical specialists per capita as compared to urban residents, and in addition significant transportation barriers to access health care. The highest death rates for children and young adults are found in the most rural counties, and rural residents see physicians less often and usually later in the course of an illness. Despite what many people think about inner-city ghettos, it's rural Americans who are at the bottom of the heap.

People in rural America experience higher rates of chronic disease, they are more likely to smoke, to lose teeth, and to be debilitated by chronic health conditions.

Americans need to ponder the question of whether a society that denies people appropriate medical care because they lack the money for it is a society that requires repair.

RACIAL AND ETHNIC DISCRIMINATION IN HEALTH CARE

IS THE AMERICAN HEALTH-CARE SYSTEM free of racial and ethnic discrimination? The answer is no. We will discuss the fallacies of racial and ethnic categorizations in Chapter 18; the issue here is whether these categorizations are the basis of discrimination in American health care, and the evidence is clear that such is the case.

As an example, consider hospital treatment in cardiac cases. Cardiac catheterization is a technique involving the passage of a small-diameter tubular instrument (catheter) into the heart through a vein or artery to withdraw samples of blood, or to measure pressures within the chambers of the heart or within the larger vessels of the heart, or to inject contrast media for visualization methods. The cardiac catheterization technique is used primarily in the diagnosis and evaluation of congenital, rheumatic, and coronary artery lesions, and to evaluate various aspects of cardiac function. It's an important and widely used technique in all clinics and hospitals in the United States that diagnose and treat cardiac patients.

Epidemiological studies have identified differences according to race and sex in the treatments of patients with cardiovascular disease, with some studies reporting that blacks and women are less likely than whites and men to undergo cardiac catheterization or coronary artery bypass graft surgery when they are admitted to the hospital for treatment of chest pain or heart attack.

In 1999, researchers confirmed the evidence that women and blacks were less likely to be referred for cardiac catheterization than men and whites, and that black women were significantly less likely to be referred for catheterization than white men. The authors of the study concluded their findings indicate that the race and sex of patients independently influence recommendations by physicians for the management of chest pain, and that decision making of physicians may be an important factor in explaining racial and gender differences in the treatment of cardiovascular disease.

More recently, studies have confirmed that racial and ethnic disparities exist in the treatment of HIV (human immunodeficiency virus) patients. The course of HIV disease is affected by the biology and social behavior of the patient, but more important is that HIV patients often have special needs and few options. Their median household income is approximately one-third that of typical households in the United States, and they are four times as likely to be members of a disadvantaged minority than to be white. Recent studies suggest that minority groups and women are underrepresented in trials of treatment for HIV infection. A report in 2002 concluded that among patients with HIV infection, participation in research trials and access to experimental treatment is influenced by race and ethnic group and the availability of patient health insurance.

But nowhere are the consequences of disparities in health care more evident than in the data concerning neonatal mortality (death at an age less than twenty-eight days). Such deaths account for approximately two-thirds of infant deaths in the United States. In 2001, blacks continued to have the highest overall neonatal mortality rate, more than twice that of any other racial or ethnic population.

In the 2000 census, 35 million people identified themselves as non-Hispanic blacks. For many health conditions, non-Hispanic blacks bear a disproportionate burden of disease, injury, death, and disability. The risk factors and incidence, morbidity, and mortality rates for the leading causes of death are greater among blacks than whites, and there are significant differences between leading causes for non-Hispanic blacks and for non-Hispanic whites, particularly with respect to homicide, HIV disease, and systemic pathogenic blood disease (septicemia). For blacks in America, health disparities can result in earlier death, decreased quality of life, loss of economic opportunity, and perceptions of injustice. For the larger society, health disparities produce less than optimal productivity, higher health-care costs, and social inequities. By 2050, an estimated 61 million black people will live in America, amounting to 15 percent of the total U.S. population. In 2050, the combined black and Hispanic populations in America will exceed the total American population of 1950.

Hispanics present a striking illustration of health disparities among American subpopulations. In the 2000 census, people who identified themselves as Hispanic constituted 12.5 percent of the U.S. population in the fifty states.

Studies indicate that for certain health conditions, Hispanics have a disproportionate burden of disease, injury, death, and disability when compared with
non-Hispanic whites, and the leading causes of death among Hispanics are different. In 2001, Hispanics lost more years of potential life before age seventy-
five than non-Hispanic whites due to the following causes of death: stroke (18
percent more), chronic liver disease and cirrhosis (62 percent more), diabetes
(41 percent more), AIDS (168 percent more), homicide (128 percent more).
In 2000, Hispanics had a higher incidence of cancers of the cervix (152 percent
higher) and stomach (63 percent higher for males and 150 percent higher for
females). During 1999–2000, Mexican Americans aged twenty to seventy-four
years reported higher rates of overweight and obesity than non-Hispanic
whites.

The consensus among epidemiologists is that socioeconomic factors (education, employment, poverty), lifestyle behaviors, social environment, and degree of access to preventive health-care services contribute to racial and ethnic
health disparities. For Hispanics in the United States, health disparities can
mean decreased quality of life and loss of economic opportunities, consequences that for society at large translate into loss of productivity, higher health-
care costs, and social inequities. By 2050, more than 100 million Hispanics will
be living in the United States, nearly a quarter of the total population, and the
consequences of poor health care will be substantially magnified.

The middle of the twenty-first century is not the far future, it's around the
corner, the time of our children and grandchildren. During that time, the combined population of blacks, Hispanics, and poor whites in America will approach 50 percent. Can this country survive as a modern industrialized society
when half its population is debilitated by inadequate health care? Conservative
economists need to ponder their support of reduced social responsibility as a
modus operandi for government and business.

IS THE CURRENT AMERICAN SYSTEM THE BEST
IN THE WORLD?

APART FROM HEALTH-CARE DISPARITIES RELATED to class, isn't it true that Americans have the best health care in the world? The answer is no, and the people
who tout American health care as in general superior to that in other countries

are twisting scientific data to suit their political agendas, twisting data and fostering a public danger.

In general, the health of a population is "public health," and public health is primarily a consequence of both culture and national policy. The United States is one of the major developed countries on the planet, with what is probably the most developed scientific enterprise, but is this intense scientific effort correlated with a comparable high level of health care for the U.S. population? For the most part, the answer is no. In addition to the fact that more than 40 million people in the United States have no health insurance, most of them with no preventive care, of the total population, 20 to 30 percent of patients who do get care receive care inappropriate for their illness—contraindicated care. And an estimated 100,000 to 200,000 people die each year from medical errors in hospitals, clinics, and physicians' offices.

Several studies have demonstrated that the U.S. population does not have anywhere near the best health in the world. In a recent comparison, of thirteen countries, the United States ranks twelfth (second from the bottom) for sixteen available health indicators. The countries in order of average ranking on these indicators (first as best) were as follows: Japan, Sweden, Canada, France, Australia, Spain, Finland, the Netherlands, the UK, Denmark, Belgium, the United States, and Germany.

On some individual indicators, the rankings of the United States were as follows. Low birth-weight percentage: 13th (last in ranking). Neonatal mortality and infant mortality overall: 11th. Life expectancy at one year for females: 11th. Life expectancy at one year for males: 12th. Life expectancy at fifteen years: females 10th, males 12th. Life expectancy at forty years: females 10th, males 9th. Life expectancy at sixty-five years: females 7th, males 7th. Life expectancy at eighty years: females 3rd, males 3rd. Age-adjusted mortality: 10th.

The World Health Organization recently confirmed the poor performance of the United States, the WHO report ranking the United States as fifteenth among twenty-five industrialized countries.

There are two common myths seeking to explain the low rankings of the United States. One myth is that the rankings are due to "bad behavior" of Americans (smoking, drinking, violence, and so on). But actual data that compares American behavior with behavior in other countries indicates behavior is not the cause of the rankings. The second myth is that the poor infant-mortality ranking is the result of high percentages of low birth-weight and infant mortality among the U.S. black population. The reality is that the international rank-ing of the United States hardly changes when data for the white population only are used.

THE PROBLEM OF MEDICAL ERRORS

ONE OF THE IMPORTANT FACTORS in a nation's overall health care is medical bungling. An argument can be made that the U.S. health-care system itself con-tributes to the nation's poor health by simple medical bungling. In the medical profession, the term "iatrogenic" generally refers to any injurious result pro-duced by medical treatment. The American record of iatrogenic consequences is not at all good: 12,000 deaths per year from unnecessary surgery. 7,000 deaths per year from medication errors in hospitals. 20,000 deaths per year from other errors in hospitals. 80,000 deaths per year from infections acquired in hospitals (nosocomial infections). 106,000 deaths per year from non-error adverse effects of medications. Iatrogenic cause is the third leading cause of death in the United States, after deaths from heart disease and cancer.

The health-care industry would like the American people to believe that better health care results from more expensive health care. The facts tell us oth-erwise, and no matter the good intentions of many hard-working socially aware clinicians, the health-care system as a whole, hugely successful as a business en-terprise, is currently a social failure.

CAN ERRORS BE REDUCED? A revealing datum about medical errors and the sta-tus of American health care is that although 71 percent of the American public believes that public reporting of medical errors is desirable, physicians are strongly opposed to the reporting of medical errors to the public. The argument that such reporting would increase the number of malpractice suits is specious: Public knowledge that hospital A has a mortality or error-injury rate ten times higher than hospital B can hardly increase the number of malpractice suits against hospital A, but it can certainly decrease that hospital's revenue. The same is true for physicians in private practice, although one doubts that much reporting of medical errors by such physicians to anyone actually exists. The problem is not malpractice litigation, the problem is that if you run medicine as a business, revealing the errors that you make gives more market share to the competition—and that's bad for business of any kind.

NOW WE ARE DOWN to the crux of the matter: a health-care system based on market forces. There are, indeed, physicians who understand the problem, as one article in a medical journal recently stated: "The current U.S. market-based [health-care] system tends to promote and offer incentives for undertreatment, nontreatment, delayed treatment, and a radical underuse of hospitalization."

But in general the situation in America is that the realities of health care are denied. Basing a health-care system on market forces assumes that an informed patient-consumer will make rational decisions about health care and that bad health care will be competitively driven out by good health care. But the assumption is a fallacy. The health care of ordinary patient-consumers is chosen by managers seeking the best deal, not necessarily the best health care, and patient-consumers who are able to make their own decisions about their health care cannot make rational decisions because they lack adequate information, especially adequate comparative information about doctors, clinics, hospitals, and other health-care providers. The present system of health care does not work, and it cannot work without a national health-care policy directed toward the public interest rather than toward the interest of an industry.

WHAT ABOUT UNIVERSAL HEALTH INSURANCE?

WOULD UNIVERSAL HEALTH INSURANCE AMELIORATE the problem of health-care disparities in the United States? Our neighbor to the north has taken the trouble to do the experiment for us. In Canada, public health insurance evolved during the second half of the twentieth century, and by the early 1970s all Canadian provinces and territories provided universal coverage for hospital and physician services. In 2001, Canada spent 9 percent of its gross domestic product on health expenditures. In the United States, the introduction of Medicare and Medicaid for only certain segments of the population demonstrated that the resulting improved access to health care substantially reduced disparities in health-care utilization.

Comparisons of physician and hospital use in the Canadian province of

Ontario and the United States have indicated that the Canadian single-payer system has been successful in redistributing services to those with the most need, namely lower-income people. Comparisons of cancer survivors in Canadian and U.S. cities have demonstrated that Canadians do better, especially those cancer patients in lower socioeconomic groups. The Canadian "experiment" indicates that universal coverage of family physician and hospital services does ameliorate socioeconomic differences in mortality, and that universal coverage of health care can be an effective means to reduce socioeconomic disparities in health. Universal coverage need not restrict the rich from buying what they like in the medical marketplace; it would merely prevents people from dying for lack of a bank account.

And what does the Canadian medical community think of the American system? A recent article provides the answer: "Canadian physicians should not be seduced by the sophisticated (but on a population basis, haphazard) 'non-system' of medical care south of the border."

HEALTH CARE AND ECONOMICS

NEVER MIND THE MEDICAL BUNGLING, for bungling is a common human frailty, or the pros and cons of various schemes, for management schemes of any kind are never perfect. What is important, what is at the heart of the problem is that the present American system of health care is inequitable—all men may be created equal, but they are not equal in the health care they receive in America, and the inequality involves money: If you can afford it, you get health care, if you can't afford it, you will probably die an early death. To be poor in America is to be expendable.

An array of studies indicate that the economic structure of a nation is the most important determinant of the health of its population. In a country like the United States, economic structure determines the availability to the masses of goods and services, particularly those goods and services necessary for personal preventive health care. People struggling at the margin of economic survival usually do not have many lifestyle choices or vocational choices. The struggle is not to stay healthy but to stay alive in a nation whose establishment derides poverty, social services, and low-cost medical care, an establishment that hurls the invective "socialized medicine" at every attempt to make health care more universal. So if 50 million people in American are living on the margin of economic survival, they constitute a huge bloc of people generally handicapped by inadequate preventive health care, a huge bloc that suffers disproportionate disease and that constitutes a disproportionate drain on the national economy. The rich don't mind it; they remain aloof from the boiling pot of daily struggle; they don't care much about the national economy as long as the national economy is used to preserve their riches.

THE CYCLE OF POVERTY, illness, and more poverty is clear to anyone who knows anything about the poor, and especially clear to physicians and social workers who practice their professions among the very poor and working poor. Poverty results in stress, and stress has debilitating effects on health. Poverty makes people sick, and once they are sick, their poverty means they are not adequately treated.

The public attitude needs to be changed. The public and the U.S. Congress need to be educated to understand that the most important contribution of medicine to society is not saving lives in acute emergencies but preventing such emergencies before they happen. Emergency medicine may be favored as dramatic material on television, but it's not the main work of medicine in a modern society.

Health care is applied science, a problem in medical engineering, and the twisting of the assessment of health care in America is certainly a form of junk science. Too many people suffer for it. The sociologist Michael Harrington, writing about poverty in America, about the "other America," closed his classic book with the following words: "For until these facts shame us, until they stir us to action, the other America will continue to exist, a monstrous example of needless suffering in the most advanced society in the world."

10

THE TALK-THERAPY FLEA MARKET

> The autonomous therapist must discover what the client wishes to buy and inform the client what he is offering to sell. . . . If you're a psychiatrist, do not let your medical training get in your way.
> —THOMAS S. SZASZ, M.D., PSYCHOANALYST

> Like politics, the art of the possible, yet also like religion, the care of souls, therapy seems true to nothing but uncertainty.
> —JOEL KOVEL, M.D., PSYCHOANALYST

HUMAN BEHAVIOR IS ONE field of science where junk science has existed since the first appearance of human intelligence. While primitive man succeeded in developing fire, tools, the beginnings of agriculture, and the use of metals, the application of reason to human behavior was hardly successful in producing an understanding of behavior "normal" or "abnormal." And the following ten thousand years did not improve matters: The transformation of the science of behavior from junk science to the beginnings of real science did not occur in force until the twentieth century, and then only in small pools of knowledge that haven't yet dribbled into the public imagination. Everyone is interested in behavior and its aberrations, the media find huge entertainment value in the subject, and whole industries are based on manipulating the views people have about their own behavior and the behavior of others. As for aberrant behavior, it's difficult to make a case that public understanding has progressed from what it was a hundred years ago. Welcome to the bazaar of modern "mental" illness.

THREE CASE HISTORIES

LET'S CONSIDER THREE CASES AND then discuss them.

The case of the patient on the wrong floor.

Michael B. is a fifty-year-old banker, intelligent, father of two grown children, an intense man of considerable vitality who has been experiencing some stress following a recent merger of his bank with a larger bank. Although the stress is not extraordinary, after a month or two, he begins to suffer vague stomach pains and nausea at intervals and particularly at night. Does he have an ulcer, the classic "executive stomach"? He visits his physician, an old gentleman, the symptoms are discussed, the stress of Michael's work, a series of routine tests are made, and everything is negative, no sign of an ulcer, no sign of anything.

The internist, who is not that fond of Michael's intense manner, does not suggest an endoscopy, since his hands have become arthritic, and in any case he thinks Michael is a typical neurotic and his condition the result of a neurosis. The internist persuades Michael that psychological stress is the likely cause of his symptoms, and they both agree that a visit to a psychologist might be useful.

Michael does see a psychologist, spends more than three years in psychotherapy, the psychologist having convinced him his stomach problem is psychological in origin. Weren't all the tests negative? Meanwhile the pain and nausea increase, and finally the psychologist suggests hospitalization in the psychiatric ward of a local hospital. So Michael enters the hospital and he's persuaded to put himself under hospital psychiatric care until his psychiatric condition is adequately investigated and cured. He arranges an extended absence from the bank and he moves in on the eighth floor of the hospital, the psychiatric ward.

Three months pass, psychological tests, psychotherapy on a continuing basis, an array of psychiatric drugs, nothing works and the pain is becoming less sporadic and more definite. Michael is becoming suspicious that his stomach pains may have nothing to do with psychological stress. He discusses the issue with the attending psychiatrists and they assure him that the most likely explanation for his problem is a psychosomatic reaction to his personal problems. He's now fifty-three, after all, the typical age for a midlife crisis. They recommend he continue with psychiatric therapy until his symptoms are relieved.

But Michael balks. He leaves the hospital, moves to the psychiatric ward (also the eighth floor) of another hospital. After a short time of more psychiatric treatment, he finally demands a complete physical examination by the gastroenterology department on the sixth floor. At last, after nearly four years of psychotherapy and psychiatric treatments, Michael receives an endoscopy and a metastasized adenocarcinoma is discovered in his stomach. This type of cancer is often curable if caught early, but at this stage the cancer is inoperable, and within another three months the cancer has spread to Michael's brain and he's dead. He dies on the oncology floor, the fifth floor. When a malpractice suit is filed against the original internist and the two hospitals, Michael's widow comes out of the final settlement with several million dollars but no husband.

THE CASE OF THE football player who went crazy.

Frank C. is a forty-year-old husband and father of two children, an easy-going fellow in a middle-management job with a large produce distributor. Twenty years ago, Frank was well-known as a halfback on the football team of a large Midwestern university. A bad knee injury in his senior year ended Frank's football career, but the glory he earned on the football field in high school and college has always been an important part of his life. He married his high school sweetheart and they've never been apart. Frank is no longer what he was physically as a college student, but if you look at him you might suspect he once played football: He stands 5'10" tall, weighs 230 pounds, and he looks extremely strong. He's been in group psychotherapy once a week for nearly a year to help with a drinking problem, and so far the therapy has been successful.

One Saturday afternoon, Frank is standing in line in a drugstore waiting to pay for a tube of toothpaste and some razor blades. He suddenly turns to one side, jerks his body, and goes berserk. He shouts at no one in particular, knocks down the young man in front of him, flails around with his arms, taking swipes at anyone who gets near him. The manager of the drugstore immediately calls the police, and within a few minutes two police officers are on the scene. They try to subdue Frank, but Frank is too powerful. He knocks the police officers on their backs. They call for help, and before long sirens converge on the drugstore and six police officers are on Frank, holding him down, people agog, merchandise scattered off the shelves, the drugstore in chaos. Frank is finally handcuffed, tied and dragged to a police van, and hauled to the nearest police station like a wild animal.

At the police station, Frank passes out, collapses on the floor in a heap, and a medical emergency crew is called. Frank is sedated and removed to the emergency room of a local hospital. An hour later he's in the neurological ward surrounded by two neurologists and half a dozen residents and interns. After several more hours, tests are completed, results discussed, and a likely diagnosis agreed upon: The patient has an abscess in a deep part of the brain, probably in a place called the "amygdala."

An abscess is an infection that produces pus, a localized inflammation that in this case may be about the size of a large pimple. If you get a pimple on your cheek, it's an annoyance that lasts a few days and then goes away. If you get the same pimple inside your brain, you may wind up killing your mother. Frank, in fact, is lucky. He gave the police a difficult afternoon, but they're accustomed to much worse. In four weeks, after successful treatment with antibiotics and some specialized oxygen therapy, Frank is out of the hospital, back at work, back to his old self as an easy-going gentleman who once had a fine few years as a celebrated football player.

THE CASE OF THE ex-marine who saw witches.

Eddie W. is thirty-six years old, divorced, and an ex-marine. Although not a former athlete, Eddie is strong, aggressive, and serious. One afternoon Eddie is pushing a cart along an aisle in a supermarket when a young man accidentally hits Eddie's cart with his own cart. Eddie curses, the young man apologizes, Eddie curses again, the young man stares at Eddie, and after a frozen moment Eddie raises his arm, chops down at the young man at the joining of neck and shoulder, and breaks the young man's collarbone.

The young man slumps to the floor in agony, a woman down the aisle sees what has happened and screams, in a few moments the store manager, two burly butchers from the meat department, and the store security guard are on the scene, overpowering Eddie, pinning him to the ground and holding him there until the police arrive.

At the police station, when Eddie is interviewed by two detectives, he insists he was defending himself, that he was about to be attacked by a Wiccan. A Wiccan? Eddie says the young man he struck at was a Wiccan, a witch, he could see it in the young man's eyes, and the young man was about to kill him. Eddie says he even heard the young man mumbling a spell before he hit him.

Eddie is kept in a lockup overnight, and the next day he's interviewed again, this time by a psychiatrist. The psychiatrist learns that Eddie has been under medication for several years, that he has recurrent hallucinations, that he often imagines people, usually witches, are talking about him, sometimes talking directly to him, that he knows they plan to kill him, and that he's extremely careful about people because the people who want to kill him are all around him. The psychiatrist also learns that Eddie has been under intermittent psychiatric treatment for several years, that he's had psychotherapy of one sort or another at various clinics, but that Eddie has never felt the psychotherapy did anything for him.

The psychiatrist finishes the interview, gently tells Eddie to relax, that at least in the police station he's safe. He offers Eddie medication, but Eddie refuses, says he won't take any drugs. The psychiatrist tells Eddie he'll write up his report and he'll see Eddie the next day. Later, when the psychiatrist is in his office, he writes his report about the patient: Paranoid schizophrenia with violent tendencies. Refuses medication. Recommend temporary hospitalization for further assessment.

Later, close to midnight in Eddie's cell, Eddie uses his belt to hang himself and end his life. The next day an autopsy is performed by a medical examiner at the morgue, with particular attention to the brain. No abnormality is found. No pimple in the brain. Eddie's brain outside, inside, pulled apart, sectioned, whatever, looks like any ordinary brain, maybe an arguable slight difference in neuron density in the frontal lobes, but nothing definitive and no apparent lesions anywhere. Medical records indicate that Eddie was diagnosed as a schizophrenic five years after he was discharged from the marines, but nowhere in his brain is there any visible sign of disease.

THE FIRST TWO CASES, the banker and the ex–football player, are based on real events; the last case, the ex-marine, is a composite construction, but a case common enough in many American jails. These three cases focus on the interface between mind and brain, between mental illness and neurological dysfunction. Only in the first case did medical professionals make serious errors, but all three cases illustrate the difficulties both professionals and ordinary people may have in understanding aberrant behavior.

An internist and two entire departments of psychiatry were convinced that Michael the banker had a "mental" illness when the cause of his symptoms was stomach cancer. Frank the ex–football player was lucky: The first medical people who encountered him did proper tests and made a proper diagnosis. In another place or another time, Frank might have been misdiagnosed and allowed to deteriorate or even die in a psychiatric ward. As for Eddie the ex-marine, the psychiatrist who interviewed him in jail had no way to predict that Eddie was suicidal: Some schizophrenics are suicidal and others are not. Medication might have been forced on Eddie, but it's not certain the medication would have prevented Eddie's suicide.

For the most part, we don't understand madness, certainly not the way a medical biologist understands liver disease. And if we don't understand madness, we understand even less about psychosocial problems, somatization, anxiety disorders, maladaptive personalities, affective disorders, sexual disorders, eating disorders, sleep disorders, drug dependence, alcoholism, or any other categorization of behavior deviating from what we say is the "norm." We don't understand aberrant behavior, which maybe is one reason why every ten or fifteen years we like to change our categorizations of aberrant behavior to be in accord with "new thinking" about what we call mental illness. The public is hardly used to the old thinking, when new thinking is already in vogue in the psychiatric community. Is it any wonder that junk science rides like a wild cowboy in the psychotherapy marketplace?

THE REALITIES OF TALK THERAPY

ALL THREE CASES, MICHAEL, FRANK, and Eddie, underwent talk therapy of some kind, either short-term or long-term. In Michael the banker's case, the talk therapy was useless because there was no reason for it—his symptoms were caused by a physical illness. In Frank the ex–football player's case, talk therapy helped him with a drinking problem. In Eddie the ex-marine's case, talk therapy over the years may have helped him cope with the personal consequences of schizophrenia, but it did nothing for the illness itself and hardly ever does in such cases.

No matter what the theory (or lack of it) behind any talk-therapy regime, talk therapy is not a medical "treatment," it's an educational process, more or less scripted depending on the therapist and the type of therapy. An enormous amount of words are constantly poured out by professionals about the theoretical foundations of this or that talk therapy, and how one form of talk therapy may be intrinsically superior to another form of talk therapy, but the fundamental reality is that controlled comparative studies reveal little difference in clinical outcome between one type of talk therapy and another. The differences between talk therapies are more important for the therapist than for the patient: Every psychotherapist must find a method of talk therapy compatible with his or her own personality.

The differences between talk therapies are essentially differences in styles of education, and therapists usually choose to work with one style or another according to their own proclivities and background. There is really nothing more going on, despite claims in the marketplace to the contrary. The theories of aberrant behavior behind various talk therapies, when such theories exist, range from the completely vacuous to speculative ideas couched in the current jargon of neuroscience. What we call "mind" is the output of the brain, and since for the more complex functions we don't understand how the brain works, we in turn don't understand how the mind works, no matter what any psychotherapist tells any patient.

Talk therapists, whether or not they have a medical degree, are educators rather than people dispensing "treatments." And as educators, the education of the educator is an important component of the efficacy of the education process—and an important component of the risk to the "patient" or "client" desiring to be educated. A medically trained psychiatrist or non–medically trained psychotherapist or crisis counselor or any sort of talk therapist with a total focus on personality dynamics and no focus at all on brain or other organ pathology as a cause of symptoms may be deadly to a patient.

The major danger to the public of current talk therapy remains that the people who promote and practice talk therapy are focused on mind rather than brain, and they usually do not have the expertise to recognize biochemical or pathogenic nervous system dysfunction as a possible cause of erratic behavior. The result is a great deal of intellectualized ballyhoo. The problem for the public is that hardly anything in this arena is regulated: Whether or not one needs a license to practice talk therapy depends on what one calls oneself rather than on what one knows or does in the office. The cowboy of junk science rides high on his horse, a wild desperado and a continuing public danger.

STRESS COUNSELING

OUR DESPERADO ASIDE, SOME FORMS of talk therapy have proven to be extremely useful, and certainly one of these is what can be called "critical incident stress debriefing," a form of crisis counseling usually applied to groups of paramedics, firefighters, and other professionals who regularly encounter traumatic events. In a typical debriefing session the counselor first provides basic information about common stress reactions such as sleeplessness, headache, irritability, flashbacks, and delusions. The members of the group are then asked to identify themselves and relate where they were during the tragic event. The discussion then turns to emotions, with each member of the group revealing what he or she was thinking during the event. According to the script, the focus then shifts to individual reactions, then a discussion of strategies for coping with stress, and finally suggestions for services that can provide additional help. At the end of this single session, participants are considered ready for reentry into the world; there is no follow-up meeting of the group.

This is obviously education rather than "treatment," and it is undoubtedly useful as long as the counselors remain focused on helping people under stress rather than on exaggerating the complexity of what they are doing in order to increase their fees. A similar brief can be made about any form of talk therapy: They are all education stratagems. Many can be useful with some but not all patients, but unfortunately nearly all therapies are prone to have their effectiveness exaggerated in a way that usually reduces their benefit to the public. And the flea market is crowded: Freudian psychoanalysis, psychoanalytic psychotherapy, Jungian analytical psychotherapy, existential psychotherapy, Rogerian therapy, gestalt therapy, Reichian and bioenergetic therapy, primal therapy, traditional group psychotherapy, encounter groups, psychodrama, transactional analysis, family therapy, somatic therapy, sex therapy, directive therapy, hypnotherapy, reality therapy, rational-emotive therapy, cognitive therapy, behavioral therapy, and so on. And these are merely the more respected forms of therapy, those that have some professional standards.

We'll consider the last two in the next section. What I want to stress is the diversity of therapeutic methods. Every one of these therapies is practiced by individual therapists in private practice who usually practice only one form of therapy with all their patients. All the practitioners are in competition with each other, all promoting their chosen therapy in the media and to their patients. But the idea that any one of these therapies is always better than any other for every patient is junk-science nonsense: There is no evidence to support that idea, and whatever evidence exists says the contrary.

COGNITIVE THERAPY AND BEHAVIORAL THERAPY

ALTHOUGH STRESS COUNSELING IS USEFUL, it's far removed from medical science. The two forms of talk therapy currently most connected to the sciences and most used in medical clinical settings are cognitive therapy and behavioral therapy, and separately or in combination they have proved to be moderately effective in altering behavior in some classical dysfunctions such as obsessive-compulsive disorder, an anxiety disorder characterized by the patient suffering intense anxiety if a particular ritual is not performed. Howard Hughes, of aviation fame, was apparently tormented by this disorder.

Of the world's population, 2 to 3 percent are believed to be afflicted with this condition, and in the past century such patients were among the greatest successes of Freudian psychoanalysis—the patients were unequivocally helped. These days, Freudian psychoanalysis, although still practiced by a minority of therapists, has essentially been tied up, gagged, and put in a closet.

In cognitive therapy, "maladaptive thoughts," such as an exaggerated sense of risk, an enhanced sense of personal responsibility for events, or excessive doubt, are cognitively (intellectually) challenged by the therapist in verbal exchange. Behavioral therapy, which differs in tactics, involves exposure to feared situations and actual prevention of compulsive behavior.

Both of these therapies are education stratagems, and both have a degree of effectiveness, especially when combined with appropriate psychiatric medications. But the mystery of the mind remains: We usually have no explanation why an individual exhibits one type of psychiatric disorder rather than another, no identification of etiology (causes), no causative agent, no sound reason for diagnosing a "disease," and the prescription of "treatment" is pragmatic rather than based on underlying science. If, rather than the estimated 3 percent, 97 percent of the people in the world exhibited obsessive-compulsive behavior, would such behavior be characterized as pathological? Without knowledge of causes and mechanisms of symptom formation, defining a disease is an intellectually risky business.

THE PROBLEMS OF MODERN PSYCHIATRY

ADDED TO THE AMBIGUITIES OF talk therapies is the awful state of modern psychiatry. In America, modern psychiatry is in a muddle. A branch of medicine not much respected by the other branches, it attracts residents in psychiatry primarily from the bottom quarter of American medical school classes, and even then only about 30 percent of psychiatric residencies are filled by American graduates, with 70 percent of psychiatric residencies filled by foreign medical graduates, mostly from developing countries, many with poor English language skills and low scores on the required board examinations. The residencies need to be filled, and hospitals and clinics take the psychiatry residents they can get. So the training situation is on the verge of being desperate.

And after training, there is the great schism between dynamic (psychodynamic) psychiatry and biological psychiatry, a schism now more than a century old and ever-widening, a split down the heart of the discipline. The reality is that in the United States most psychiatric psychotherapists (psychiatrists who primarily do talk therapy) are limited to a primitive knowledge of psychiatric drugs and primitive knowledge of modern neuroscience, while prescription-writing "biological" psychiatrists who do not do talk therapy are known to have an uncertain command of general medicine and are often too unfamiliar with the psychodynamic problems of their patients. As psychiatrists J. A. Hobson and J. A. Leonard point out: "Very much like a victim of psychosis, psychiatry has lost much of its ability to relate to the real world and do its job."

Do the members of the medical psychiatric community understand the magnitude of their current problem? Some certainly do. The psychiatrist E. Fuller Torrey has stated the problem in an engaging simile: "Psychiatry is an emperor standing naked in his new clothes. It has worked and striven for seventy years to become an emperor, a full brother with the other medical specialties. And now it stands there resplendent in its finery. But it does not have any clothes on; and even worse, nobody has told it so."

And outside the castle of psychiatry, the Goths are seeking ways to cross the moat: Not only are clinical psychologists demanding the right to prescribe psychiatric drugs, but general medical practitioners have become leading prescribers of Prozac and other antidepressant medicines. The drug companies, in fact, do not need to depend on psychiatrists for their revenues: Intense promotion by drug companies to general practitioners for on-label and off-label use of psychiatric drugs is now ordinary in American medicine.

Is it safe? Sometimes yes and sometimes no, and most often the public serves as the sorry guinea pig to assess side effects. But whether or not the prescription of psychiatric drugs by general practitioners (and potentially by psychologists) is apparently safe at the moment, the prescription of a psychiatric drug by a general practitioner with only a primitive or outdated understanding of psychiatry and brain science is nothing but the practice of junk medicine. Such prescribing is common, the damage to patients when it occurs is largely unreported, and the public is once again at the mercy of the piper in the three-cornered hat: physician, pharmacist, and pharma. Truly, to ask the question, Is it safe? about a system of medicine based on market forces is not sensible: Such a system cares more about profits than safety, and prescribing and selling Prozac, no matter what the collateral damage of side effects, is certainly a profitable enterprise.

THE DIAGNOSTIC ZOO BOOK

CLINICAL PSYCHOLOGISTS, WHOSE INTELLECTUAL FOCUS is usually diagnostics rather than treatment, also have their quandaries as they struggle to navigate the labyrinthine maze of the current *Diagnostic and Statistical Manual of Mental Disorders* (called DSM), the compulsory diagnostics guide for practitioners in the mental health professions, the diagnostics manual required by insurance companies for reimbursement, and therefore the single most important item in print for the practicing talk therapist seeking to earn a living as a therapist. The manual, which is now in its fourth version (DSM-IV) consists of nine hundred pages defining nearly three hundred mental illnesses. It's the working bible of psychiatry, clinical psychology, the justice system, social-service agencies, schools, prisons, and governments. The problem is that no one is certain the DSM is anything more than a convenient and changeable classification scheme that benefits professionals more than patients, if it benefits patients at all.

The manual states: "In DSM-IV, there is no assumption that each category of mental disorder is a completely discrete entity with absolute boundaries dividing it from other mental disorders or from no mental disorder."

Fine. But then on the same page, one finds: "DSM-IV is a categorical classification that divides mental disorders into types based on criteria sets with defining features."

The first statement, essentially a disclaimer, is usually ignored by people who use the manual. The question of whether various mental disorders are discrete clinical entities or arbitrary distinctions in academic psychology has been debated for more than fifty years, but it's not the primary concern of the people who use the manual. Their concern is the second statement, the manual as a convenient categorization of nearly 300 mental disorders, labels set down on paper with the blessing of the medical community via the American Psychiatric Association, a labeling scheme that has become essential to running the business of mental health in America. Never mind intellectual debates about fundamentals, we're practical people here and we need organization!

If this universally used manual is indeed the detailing of an arbitrary classification scheme, it immediately becomes a junk-science primer, and perhaps in the twenty second century it will be recognized as such the way we recognize today as junk science the diagnostic scheme of nineteenth-century phrenologists who diagnosed mental illness by consulting an atlas of bumps on the skull. Having insurance companies and government bureaucrats decide the structure of a science in the form of a diagnostic manual is uncomfortably close to madness itself. Again, the desperado rides high.

THE END OF PSYCHIATRY

MANY PEOPLE, AND I AM one of them, believe that the end of psychiatry will occur in this century. Neurology treats diseases of the brain; psychiatry treats diseases of the mind. No matter that there is no such thing as mind without brain, that "mind" is produced by the brain, that the brain is most definitely the organ of all behavior, the intellectual and medical dichotomy is institutionalized. The cut, indeed, is analogous to a division between diseases of the gastrointestinal tract and diseases of "digestion," except that on scientific grounds the latter division, which does not exist in any formal sense, would be considered medically irrational. Ultimately, as "biological" psychiatry becomes more and more fused with neuroscience and neurology, the faculty of departments of psychiatry will be merged into the faculties of medical neuroscience and psychiatry will cease to exist as an independent medical discipline. Academic departments, after all, are artificial man-made categorizations of knowledge, categorizations that serve either as ways to get funding or ways to organize teaching. My view is that psychiatry departments will probably disappear by mid-century.

As for talk therapy, particularly private-practice talk therapy, I believe it will have a robust future until one day people will realize that a sophisticated "counseling machine," appropriately programmed to interact with a user, can provide most of the personal education currently offered by various forms of psychotherapy. This will be particularly true of the more scripted psychotherapies, therapy regimens in which the therapist is already acting as a pseudo-machine. Counseling machines will in fact make talk therapy more objective, more easily defined for research on comparative methods, and certainly more easily subjected to ethical codes. The idea of counseling machines sounds like a science-fiction prop, but not long ago that was exactly the attitude toward the idea of an interactive World Wide Web.

PART FOUR

POISON AND BOMBS IN THE GREENHOUSE

11

POLLUTION: PRIVATE INTEREST AND PUBLIC POISON

> When an executive decides to take action for reasons of social responsibility, he is taking money from someone else— from the stockholders in the form of lower earnings or from the consumer in the form of higher prices.
> —MILTON FRIEDMAN, NOBEL LAUREATE ECONOMIST, 1973

> Companies have to be socially responsible or the shareholder pays eventually.
> —WARREN SHAW, CEO OF CHANCELLOR LGT ASSET MANAGEMENT, 1997

MOST INDUSTRIES PRODUCE WASTE products that can pollute the air and the ground, or use materials in factories that can harm workers, and if we're to have an industrialized society, such consequences of industrialization are unavoidable. What is avoidable, however, is fooling the public into believing that pollution or potential harm is absent when it exists, hiding the dangers when science says the dangers may be considerable, and cutting corners on environmental and worker protection in order to maximize shareholder profits.

The public problems concerning environmental hazards produced by industry are essentially problems of combatting the junk science about pollution broadcast by industry and government, and the problem of defining and accepting the idea of corporate social responsibility.

Milton Friedman's idea, quoted above, that social responsibility costs money, is what I call "wimpy thinking"—thinking weak and ineffective in logic, thinking with a superficial attitude often attached to an even more superficial philosophical precursor. Such thinking in the hands of many executives in industry usually produces a posture that results in either hiding science or twisting science to achieve corporate ends.

A human society is a collection of individuals organized for mutual benefit but primarily for survival. When social interactions threaten survival of society, the main reason for the existence of society vanishes. The important conflict, apparently unrecognized by Friedman and others, is not between social responsibility and profits, but between short-term profits and long-term losses. In the complete absence of social responsibility, society dies (including shareholders!) and there are no profits for anyone. The idea that the need for corporate social responsibility is a nuisance invented by activists is maybe the most dangerous idea in the ranks of industry in America and elsewhere.

The social sciences have a term for individuals who refuse social responsibility: They are called "free-riders."

But do "conservative" executives, economists, and politicians really believe that social responsibility is a waste of money? Apparently they do, all of it wimpy thinking, since their so-called conservatism does not extend to conserving the society that makes industry, business, and capital gain possible.

At least some of these people must understand they cannot kill society and still live in it. Maybe the general idea is to bleed society slowly, not enough to kill it, only enough to keep the profits rolling in while keeping the weakening creature alive long enough to at least secure the privileges of one's grandchildren. But as a business strategy, this is certainly too dangerous—both for one's grandchildren and for the rest of society.

The image evoked is vampirism. But the classic vampire bleeds his victim with restraint in order to prevent the victim's death and the consequent loss of blood supply. Too often our industrial vampire seems confused about how much restraint is necessary, the result a stupid vampirism that sometimes kills the vampire himself after social discovery and enormous financial losses.

Meanwhile, ordinary people die, collateral damage in the rush for profits. Some die quickly and some die slowly, but they die like poisoned flies.

INDUSTRIAL POLLUTION

IS INDUSTRIAL POLLUTION TOTALLY AVOIDABLE? There is surely no way to have an industrialized human society without a degree of human error or ignorance of danger and consequent environmental pollution. To be completely free of such consequences, we would need to revert to the Old Stone Age, or even earlier to some age before fire, since burning fires inside poorly ventilated caves must have on occasion caused carbon monoxide poisoning. Error and ignorance of consequences have been with us since the beginning of human history, and they are not leaving us any time soon.

Our focus here, however, is not mistakes or ignorance that can cause catastrophes such as Love Canal or Chernobyl or Bhopal. Our focus is the twisting of environmental and medical science to hide pollution and cause public harm—twisting often with the collusion of government, media, think tanks, a cabal of wimpy thinkers who believe society is the enemy and social responsibility a nuisance idea invented by a radical mob.

LEAD POLLUTION

THE HUMAN NERVOUS SYSTEM DOES not like lead. This metal, in trace quantities as the doubly valent lead ion, poisons the nervous system, probably by substituting for the doubly valent calcium ion, and is especially harmful to the fetus, the infant, and the young child—in all of whom the nervous system is developing at a frantic pace. Most of the overt toxicity from lead results from environmental and industrial exposure: leaded gasoline, leaded paint, and lead in drinking water.

Beginning in 1925, tetraethyl lead was introduced as a gasoline additive to improve fuel-burning efficiency. Tetraethyl lead is not "liquid lead," which is molten metal. Tetraethyl lead is an organic lead compound used as an additive in liquid gasoline. The toxicity of lead for humans was well-known, but the gasoline manufacturers, who called the new gasoline "ethyl," insisted the lead additive was present in quantities too small to be harmful to the public. When in 1924 workers at a Standard Oil ethyl plant in New Jersey began suffering from acute neurological effects of lead poisoning after gasoline spills, Standard Oil management brushed it off: "These men probably went insane because they worked too hard." Seventeen workers at the plant died.

At a public hearing in 1925, Standard Oil executive Frank Howard declared tetraethyl lead "essential to our civilization . . . a gift of God." The official panel concluded "there are at present no good grounds for prohibiting the use of ethyl gasoline."

By 1936, 90 percent of the gasoline used in the United States contained tetraethyl lead.

Throughout its history, the Ethyl Gasoline Corporation, an offshoot of Standard Oil, General Motors, and DuPont, ran media and political campaigns to promote leaded gasoline. Experts hired by the company argued that keeping the lead in gasoline was essential for the growth of American business. Executives of the company ignored contrary advice: Where was the proof that low levels of lead were any threat to human health? Frank Howard of Standard Oil said, "Because some animals die and some do not die in some experiments, shall we give this thing up entirely?"

Lead toxicity in minute concentrations in the human body was poorly understood in the 1920s. Knowing that something is poisonous tells you nothing about how or why or how much is actually dangerous. So it makes no sense to fault either government or industry for the introduction of leaded gasoline into the industrialized world. What is important in this context are not the mistakes made from ignorance in the 1920s, but the refusal to acknowledge experimental data for so many decades afterward, decades during which countless children exposed to leaded gasoline fumes suffered serious irreversible damage to their nervous systems, damage that produced tragic mental disabilities — damage that could have been avoided.

The record of the gasoline industry is unpleasant. In the mid-1950s, U.S. automobile makers apparently agreed among themselves to fight imposition of emission-control standards by the government. In 1959, under pressure, the U.S. Public Health Service approved the Ethyl Corporation request to increase lead in gasoline. In 1969, U.S. automakers settled a suit by the Justice Department for conspiracy to delay the development of pollution-control devices. In 1973, the U.S. Environmental Protection Agency promulgated the phase-out of lead in gasoline. When the standards were set, the Ethyl Corporation sued the Environmental Protection Agency. In 1980, the Ethyl Corporation reported it had expanded its overseas business tenfold between 1964 and 1981. In 1981, a task force headed by Vice President George Bush on "Regulatory Relief" unfortunately proposed to relax or eliminate the U.S. leaded gas phase-out.

How many innocent lives were affected by the decades of delay? The primary phase-out of leaded gasoline in the United States was not completed until 1986. The European Union finally banned leaded gasoline in the year 2000. Leaded gasoline is still manufactured and sold by oil companies and is used extensively in undeveloped countries. It seems gasoline producers are guilty in the dock, and the problem of leaded gasoline is not yet solved.

Is the science clear? Lead is certainly neurotoxic, and young children are at particular risk for exposure. Many studies indicate that blood lead concentrations above 10 micrograms per deciliter are associated with adverse outcomes on measures of intellectual functioning and social behavioral conduct. Such studies led to the identification of a blood lead concentration of 10 micrograms per deciliter or higher as a "level of concern" by the Centers for Disease Control and Prevention (CDC) and the World Health Organization (WHO). But it's not clear whether lead-associated cognitive deficits occur at concentrations below 10 micrograms per deciliter. No evidence of an apparent threshold exists for lead-associated deficits, and there has been very little research on the possible effects of blood lead concentrations below 10 micrograms per deciliter. Some studies in which the average blood lead concentration was below 10 micrograms per deciliter have reported associations between the blood lead concentration and cognitive deficits, but these studies did not focus specifically on children whose concentrations remained below 10 micrograms per deciliter throughout life.

There is also some evidence of environmental-lead toxicity in adults. Much of the lead taken into the body is incorporated into bone where it constantly interchanges with other tissues. Studies suggest that accumulated lead exposure is related to several chronic disorders of aging, including hypertension and cognitive decline, disorders that have been associated with oxidative stress. Several lines of evidence suggest that accumulated lead exposure could also increase the risk of another oxidative-stress-related disorder of aging, age-related cataract—the leading cause of blindness and visual impairment worldwide.

AIR POLLUTION

UNFORTUNATELY, LEAD IS NOT THE only toxic contaminant in the air we breathe, and efforts by industry and government to do battle against public concerns about other contaminants have been just as vigorous as the efforts of gasoline companies to reduce concerns about lead pollution.

What evidence is there concerning the toxic effects of air pollution in general? Extreme air-pollution disasters in the past have kept the public aware of dangers: Meuse Valley, Belgium, in 1930; Donora, Pennsylvania, in 1948; London in 1952. These and other fog and smog disasters have provided evidence of episode-related increases in morbidity and mortality from cardiopulmonary damage, demonstrating that extremely high concentrations of air pollution can have serious negative effects on health. Such disaster episodes provoked efforts to improve air quality in the United States, UK, and elsewhere, and extremely severe episodes of air pollution were largely eliminated in many industrialized countries.

The term "particulate-matter air pollution" refers to air pollution that involves small solid particles or liquid droplets suspended in air: soot, fumes, dust, pollen, spores, smoke, and so on, the class of suspended particles commonly found in cities and industrial areas. In the early 1990s, several new epidemiological studies reported adverse effects of unexpectedly low levels of particulate-matter air pollution, and these data prompted reconsideration of particulate-matter standards and health guidelines. In general, research has continued to suggest that a level of air pollution common in many urban and industrial environments is an important risk factor for various adverse effects in humans. Many of these studies have focused on respiratory disease, but substantial evidence indicates such pollution is also a risk factor for cardiovascular disease. Short-term exposure can be dangerous for anyone with existing pulmonary or cardiovascular disease, and long-term repeated exposure to particulate-matter air pollution produces a cumulative risk of chronic pulmonary and cardiovascular disease and death.

In recent studies, particulate-matter air pollution has also been implicated in mutagenesis in animals, and the possibility that such effects also occur in humans is troubling. By monitoring changes in DNA sequences, researchers have demonstrated that offspring of mice exposed to an industrial location on western Lake Ontario have an increased rate of apparent mutations and these genetic changes are paternally derived. The researchers discovered that the mutation rate could be reduced by approximately 50 percent by cleansing the air with a high-efficiency-particulate-air filter, and this suggests that particle-bound mutagens, or the particles themselves, are responsible for the observed heritable DNA changes.

There is also mounting evidence that air pollution has chronic adverse effects on pulmonary development in children. Studies conducted in Europe and the United States have demonstrated that exposure to air pollution is associated with reductions in the growth of lung function. Previous studies had followed young children for relatively short periods (two to four years), leaving unresolved the question of whether the effects of air pollution persist from adolescence into adulthood. New research indicates that current levels of air pollution have chronic adverse effects on lung development in children from the age of ten to eighteen years, leading to clinically significant deficits as children reach adulthood.

The idea that the air we breathe can be harmful is disturbing, since air is vital to our existence, something we take for granted, something without which we cannot survive. The assurances of various industries that pour waste into the atmosphere, particularly the assurances of the chemical industries, are not enough. Such assurances are too often based on a desire for short-term profits coupled with a disdain for those with a focus on social responsibility. These attitudes are unfortunate and boorish and must be condemned.

Constant surveillance in the public interest is needed, and history demonstrates that the only entity capable of such sustained extensive surveillance is the government. Surveillance and continuing experimentation to delineate and understand harmful effects of air pollutants are necessary, and both are expensive. But the cost to society of poisoned air is even more expensive and ultimately unbearable.

XENOBIOTICS

THE TERM "XENOBIOTICS" REFERS TO molecules foreign to biological systems, and here on Earth such molecules are for the most part synthetic compounds produced by industries, particularly chemical industries. During the past fifty years, the rapid expansion of chemical industries in both the developed and developing worlds has resulted in the release of a variety of xenobiotics into the environment. These chemicals have worked their way into our lives in a variety of forms: pesticides, herbicides, cosmetics, preservatives, cleaning materials, municipal and private waste, pharmaceuticals, and industrial by-products. Awareness of the biological risks of chemical toxicity has increased considerably in recent years, but some of these chemicals have long half-lives and have been detected in environmental samples ten to twenty years after they were banned for sale or use.

Analysis of the biological consequences of environmental pollution has generally centered on the risks for induction of certain kinds of cancer. But it's becoming increasingly apparent that another major target of environmental chemical pollution is the reproductive system, particularly the male reproductive system. This was first recognized more than thirty years ago, when male workers exposed to 1, 2-dibromo-3-chloropropane, an agricultural control agent used to kill nematode worms in soil, developed severe disruption of sperm development and associated infertility. Since then, numerous independent studies have related occupational exposure to pesticides, herbicides, industrial agents, and heavy metals to poor semen quality and impaired male fertility.

Many companies are involved in the production and distribution of xenobiotics for one use or another, chemicals that for the most part are of great utility in our lives. To demand that every chemical produced be guaranteed to be completely harmless is not a useful demand, since nothing can be guaranteed to be completely harmless. Instead, it might be useful to society to improve the education of chemical company executives, improve their education in biological sciences, improve their understanding of the chemical frailty of biological systems. Most chemists and chemical industry executives have had little education in biology, and they need to be made aware of how their own children and grandchildren may be at risk when xenobiotics are produced and released without any consideration of possible consequences. A consciousness about the need for caution would be a good starting point for reducing public hazards.

The executives of chemical companies could be taught enough biology to understand that for any biological system there is no such thing as a "toxicity threshold." Such a threshold always depends on our current knowledge of a system and on the instrumentation available at the time we make the determination. Tomorrow, both knowledge and instruments may be different, and we may discover too late that our "threshold" was too high.

Biological systems have no real lower limit of vulnerability. A single alpha particle can cause a serious mutation in the genome. A single molecule of a xenobiotic at a crucial place in the genome can in principle cause enough of a disruption of gene regulation to produce a cancerous cell. There are no lower limits because in biological systems it is organization that rules and not mere numbers of molecules. Disrupt organization at critical points and havoc may result.

Chemical company executives need to be taught that they, their spouses, their children and grandchildren are vulnerable to every xenobiotic that enters the body. Maybe then the attitudes of the chemical industry would be altered; if not, then we are all truly in the hands of a collection of witting and unwitting social vampires.

MERCURY

MERCURY IS A SPECIAL CASE. The toxicity of mercury has been known for at least 2,000 years. In the first century A.D., the Roman historian Pliny the Elder (23–79 A.D.) described the mercury poisoning of slaves who worked mercury mines contaminated with mercury vapor, work considered too unhealthy for Roman citizens. There is no possible excuse in the case of mercury; there is no excuse for any entity, commercial or otherwise, to release mercury into the environment in a way that may be harmful to humans. Any attempt by industry or government to belittle the dangers of mercury pollution is at best a dangerous twisting of science against the public interest.

Mercury is currently used in the production of chloralkalies (e.g., bleach), electrical equipment, various paints, thermometers, dentistry, and in various research laboratories. Most human intake of mercury vapor is from dental amalgams; most human intake of methyl mercury is from fish; most human intake of inorganic mercury salts is from food other than fish.

Mercuric nitrate poisoning was a common occupational hazard in the felt-hat industry in the sixteenth century (mercuric nitrate was used in the processing of felt fabric). Mercuric salts are widely used in industry today, and industrial discharges of mercury compounds into rivers have contaminated the environment in many parts of the world. Alkylmercury salts, especially methyl mercury, are the most dangerous organomercurials. Major incidents of mercury poisoning from human consumption of mercury-treated seed grain have occurred in various undeveloped countries. In the fall of 1971, Iraq imported large quantities of wheat and barley seed treated with methyl mercury, and the grain was distributed for spring planting. Ignoring government warnings, people ground the grain into flour and used the flour to make bread. The consequent catastrophe hospitalized 6,500 people, 500 of whom died.

The disease we know of as "Minamata disease" involves methyl mercury poisoning. Minamata is a small town in Japan whose major industry was a chemical plant that emptied industrial waste directly into Minamata Bay. The chemical plant used inorganic mercury as a reaction catalyst, and some of this mercury became methylated (probably by microorganisms) before or after discharge into the sea. The compound apparently was taken up by ocean plankton algae and became concentrated in ocean fish consumed in large quantities as a staple by the residents of Minamata. These residents were the first to be poisoned, with forty-six deaths and tragic consequences for many who survived. In the United States, human mercury poisonings have resulted from eating meat from pigs fed grain treated with an organomercurial fungicide.

Epidemiological studies have in general recognized neurodevelopmental effects in young children from in utero methyl mercury exposure in populations in which fish or seafood is a substantial component of the diet and in which exposure levels may be comparable with those levels in high-end consumers in the United States. Data indicate that approximately 8 percent of women of child-bearing age in the United States have in their bodies concentrations of mercury higher than the currently accepted toxicity threshold, and women who are pregnant or who intend to become pregnant need to follow federal and state advisories on consumption of fish.

WOMEN AND ORGANIC SOLVENTS

BIOLOGICAL MEMBRANES ARE ESSENTIALLY LIPID barriers with a secondary population of attached or embedded proteins. Organic solvents are particularly potent in their chemical action on biological systems because they readily and quickly pass through lipid domains such as biological membranes, including membranes of the skin, the digestive tract, and the respiratory system.

Many women of child-bearing age are occupationally exposed to organic solvents, and since fetal tissue is even more vulnerable to such solvents than adult tissue, the question of the effects on the fetus of occupational exposure of the mother to organic solvents is of some significance.

Two women-dominated occupations with potential chemical exposures are health-care professions and work tasks in the clothing and textile industries, all of which involve exposures to organic solvents. Many industrial solvents are teratogenic (i.e., capable of producing a malformed fetus) in laboratory animals, and there are reports of limb and central nervous system defects in mice, marked developmental toxic effects and retardation of skeletal growth in rats, and congenital malformations in rabbits.

However, the animal studies typically use high doses of single solvents and a variety of routes of administration. In the human occupational setting, exposure usually occurs to a multitude of solvents at much lower doses by inhalation, making extrapolations from animals to humans problematic.

In studies in humans, significantly more major malformations occur among fetuses of women occupationally exposed to organic solvents than among matched controls. The numbers are clear: In one study, 10 percent of the women in the organic-solvent-exposed group produced a malformed fetus, as opposed to less than 1 percent of the women in the control group. At the present time, only one conclusion is reasonable: Occupational exposure to organic solvents during pregnancy is associated with an increased risk of major fetal malformations.

These studies of female workers were not experiments, they were simply studies of present-day conditions and consequences in American industries. Whatever occupational safety regulations are in force are obviously of no benefit to many women exposed to indoor pollution hazards. But rather than improve safety regulations, the push of the Bush administration has been to remove them. Government bureaucrats, many of whom are former lobbyists for the industries they were appointed to regulate, continually produce reports minimizing occupational health problems, reports based on twisted science and skewed data. While corporate executives spend balmy afternoons on the golf course, and corporate shareholders smile as they look at their dividend checks, women who work in factories worry about how much poison they can take before having babies with two heads.

THE HIGH-VOLTAGE TRANSMISSION LINE FIASCO

ONE OF THE CURRENT PROBLEMS with attempts to protect society from environmental pollution is that environmental activist groups are often too prone to attack industry and government without appropriate understanding of scientific realities. If we are to insist that industries do their homework before releasing contaminants into the environment, it behooves those who attack various industries to do their homework as well. Absent this, we are in a circus that benefits no one except the circus-loving media producers.

During the 1980s and 1990s, the public was bamboozled by activist groups and some scientists into believing that high-voltage transmission lines, ubiquitous in the United States and in other industrialized countries, might cause cancer, especially leukemia. There is no question that strong electromagnetic fields can cause disruptive changes in biological cells, but most biologists and physicists have always failed to understand how the extremely weak electromagnetic fields associated with transmission lines could possibly produce pathological processes. The scientific consensus, the consensus of biologists and physicists completely independent of industry or government, has always been that the fields are too weak and the activism unwarranted.

Finally, in 1997, a major study appeared, a report describing the results of the most extensive and carefully controlled observations to date, the report concluding that the supposed relationship between electromagnetic fields and leukemia does not exist.

Two years later, in 1999, the U.S. Office of Research Integrity reported the intentional falsification and fabrication of data during the 1990s by a biochemist studying the effects of ambient electromagnetic fields (power lines and home appliances) on biological cells. The researcher, Robert P. Liburdy, who worked at the Lawrence Berkeley National Laboratory, agreed to retract papers published as far back as 1992, papers that played a significant role in the public brouhaha about high-voltage transmission lines as a cause of leukemia. Liburdy, in fact, had been awarded more than $3.3 million for his research on the biological effects of electromagnetic fields.

Here, then, is a case where both junk science produced by ignoring science and junk science produced by scientific fraud coalesced against the public interest. More than one medical scientist has pointed out that the large amount of money wasted on electromagnetic-field-leukemia studies could have been used for basic research in this deadly disease that produces so much tragedy in children and adults.

PROSPECTS FOR PUBLIC SAFETY

IN AN IDEAL SOCIETY, DECISIONS about public safety would be made on the basis of evidence and without bias determined by special interest pressures. The U.S. Environmental Protection Agency describes its function as follows:

> Ideally, regulators would like to eliminate all pollution and its risks, but this is usually not a realistic expectation. Regulators must address the most important risks and decrease them to the level at which they believe the risks are smaller than the benefits of the activity causing the pollution.

The problem is that the people who make decisions about risks and benefits are for the most part political appointees, changing with the winds of American politics, and during their time in office subjected to pressures from all sides, and in particular to pressures from private industry, the main source of hazardous air and water pollution in the United States. Science is continually pulled, pushed, twisted, and shredded in the attempts to protect profits against the public interest. Until industry executives themselves are imbued with a desire to do no harm and with an acute awareness of how easy it is to cause chronic human misery with industrial pollutants, not much will really happen to protect the public.

Meanwhile, we need public vigilance against junk science emanating from the chemical, textile, fuel, and other industries, and from government agencies that should be protecting the people at large.

12

MISSILES AND TERRORISM: NO PLACE TO HIDE

> Much science is done in the academy, where the poisonous vapors of campus claptrap waft even into the physics lab.
> —JOHN DERBYSHIRE, *NATIONAL REVIEW*

> A nuclear explosion at the World Trade Center, involving two grapefruit-sized lumps of enriched uranium, would have devastated three square miles of southern Manhattan, including the whole of Wall Street.
> —MARTIN REES, PHYSICIST

ARE WE SAFE? The question has been at the forefront of American politics for more than sixty years, ever since the beginning of the atomic age. But given the methods of modern warfare, and more recently the methods of terrorism, the question is now more a question for scientists rather than a question for politicians and generals. Unfortunately, politicians and generals don't always welcome the answers of scientists, and when that happens, the answers of scientists are likely to be kept secret or to be buried in a mountain of paper to keep the answers from public awareness. The government then practices a brand of junk science by hiding useful and important science from the public.

If it's true that science is used in war and terror, it's also true that war and terror are not made by scientists. War is a deadly tribal psychosis, a total corruption of reason, and most scientists find both the theory and practice of war repulsive. Terror is a product of fanaticism, a product of an attitude that innocent unengaged lives are expendable in the pursuit of some political or philosophical or religious belief. There is no sanctity of life or reason for the terrorist, only a sanctity of creed. As a group, scientists find terrorism as repulsive as war.

At the same time, scientists are citizens, and in time of war, governments turn to scientists for their knowledge of the real world. This, of course, leaves scientists in a state of confused ambivalence: They hate war, but in time of war they are called upon to use their brains to achieve victory. So, in the last century, scientists gave their governments atomic bombs and hydrogen bombs, and then later, predictably, they repented and demanded these weapons be destroyed to protect humanity. Unfortunately, politicians and generals are not scientists, and if it seems reasonable to scientists to bury weapons of mass destruction in a time of peace, to politicians and generals such burial is not feasible, not sensible, and not likely.

The number of living people who were alive in 1945 grows less with each passing day, and as that happens human memory fades into written history. Sixty years have passed since Hiroshima and Nagasaki. We're still here, nuclear weapons are still with us and even more powerful, and the problem of how to prevent our self-destruction is not yet solved.

MISSILE DEFENSE STRATEGIES

THE BUSH ADMINISTRATION HAS REVIVED old ideas of missile defense, and although these ideas have been opposed by many scientists as unreasonable, the views of these scientists have hardly been publicized by the government and are largely hidden from the public. This may be politically expedient for the Bush administration, but it's not in the public interest, and it leaves one wondering if this is a case of politics preempting national security.

The major premise of missile defense strategies is that a nuclear attack against the United States would involve one or more intercontinental ballistic missiles carrying nuclear warheads. That premise may have made sense during the Cold War with the Soviet Union, but as we'll see, it makes less sense now when the major threat to the United States is from international terrorist groups.

Let's review the three possibilities for a nuclear missile attack upon the United States.

One possibility is a sophisticated attack with many missiles carrying nuclear warheads, an attack that could come only from an advanced nation-state. At the present time, there are no such advanced nations threatening us with such attacks, so such an attack is unlikely. (If such an attack did occur, no complete defense is possible. Sophistication means decoys would be used, and the use of decoys means that at least some missiles would get through any defense.)

Another possibility is an attack by a "rogue" state rather than by an advanced state. The so-called rogue states of Iran and North Korea may ultimately develop nuclear missiles, but it's doubtful if these two countries will ever have the technological capacity to put up hundreds of missiles and completely destroy the United States. And failing such destruction, a feeble one-, two-, or three-missile attack would be suicidal on their part, resulting in the complete destruction of their nation-states. Such an attack is therefore deterred and unlikely.

A third possibility concerns terrorist groups. Could terrorist groups make and deliver hydrogen bombs or atomic bombs in warheads?

Nuclear warheads are not simple. A modern thermonuclear warhead (hydrogen bomb or fusion device) consists of two main elements, the primary and secondary stages. In the primary stage, chemical high explosive is used to compress a core containing plutonium-239 into a state of nuclear supercriticality, the state that provokes a chain reaction. The subsequent escalating fission process results in temperatures and pressures that allow the energy generation, or "yield," to be augmented by the fusion of a deuterium-tritium mixture—a process known as "boosting." The exploding primary stage releases copious X rays, which can then be used to implode a secondary stage with immense force. It's from the fissionable and fusionable materials that constitute the secondary stage that the bulk of the overall hydrogen-bomb warhead yield—the energy generated—is derived.

As for the less complicated atomic bomb (fission device), the required technical expertise is almost as great as that for the hydrogen bomb. The engineering of the implosion chamber necessary for the initiation of the explosive chain reaction requires considerable technological sophistication, and like the hydrogen bomb, the engineering cannot be validated without a test—an unlikely project for a stateless terrorist group.

So to have a nuclear device, a terrorist group would need to buy one or steal it, and to mount the device on a missile as a warhead, the terrorist group would need to acquire such a missile and the technical expertise to do the mounting, let alone the necessary technological sophistication to fire the missile accurately at a target. It's all possible, of course, but not the most likely scenario.

Nevertheless, because international terrorists have no nation-states to preserve, they are the most serious threat. The strategy of terrorist groups is to pro-

duce chaos, not to win wars between nations, and to achieve chaos with a nuclear weapon, missiles are unnecessary and unwieldy, and the technology is too
forbidding for a terrorist group. The more relevant nuclear strategy for a terrorist group is to obtain a single nuclear device, and instead of mounting the
weapon on a missile, mount it on a truck or ship entering a city like New York,
or simpler still, set up the device in an ordinary apartment in Brooklyn and explode it at will. The real nuclear threat is not from incoming rockets, but from
more mundane delivery methods that involve initial transport of nuclear material and hardware by land and sea. Watching the sky does nothing for defense
of the United States against nuclear terrorism.

Let's repeat the important conclusions:

1. At present, there is no reason to expect a sophisticated nuclear
missile attack, an attack that would have to come from an advanced nation.

2. If a nuclear missile attack is by a rogue state and the attack is not
sophisticated, the rogue state will be able to damage but not destroy us, and our retaliation would certainly destroy that rogue
state quickly. The threat of complete destruction makes an initial rogue-state attack unlikely. The leaders of rogue states are
not known to be suicidal.

3. A terrorist attack with a nuclear weapon is a real possibility, but
it's unlikely that any missiles would be involved.

In summary: Our current nuclear problem is terrorism, and not a war between countries, and given that, the problem is not missiles, but penetration by
land or sea with a single nuclear device or a "dirty" bomb consisting of radioactive material packed with a high explosive. A dirty bomb might not kill a large
number of people, but the consequent radioactive contamination could easily
make a modern city uninhabitable.

So where is the junk science? The answer is that it's in the hiding of what
you need to know. Has the government made a point of broadcasting the conclusions of scientists that contradict government policy? Has the media both-
ered to inform you that thousands of American physicists are in opposition to
government current policy not on political grounds but on grounds of scientific

unfeasibility? Is it proper in a democracy to hide reality from the public? Let's now consider some of the views of American physicists in detail.

HOW THE PHYSICISTS VIEW MISSILE DEFENSE

IN THE FALL OF THE year 2000, the American Physical Society appointed an advisory committee to study the possible choices concerning missile defense. This committee reached the following conclusions: Defending the fifty American states against liquid-propellant ICBMs from North Korea may be feasible, but would push the limits of what is possible physically, technically, and operationally. Defending the fifty states against liquid-propellant ICBMs from Iran would be much more challenging. Defending the fifty states against solid-propellant ICBMs from North Korea or Iran is unlikely to be practical when all factors are considered.

Defending only the West Coast against ICBMs from North Korea would be easier than defending all fifty states. Defending only part of the United States against ICBMs from Iran would not be easier than defending all fifty states.

The idea of X-ray lasers as missile defense weapons was first proposed by the Livermore Laboratory and the physicist Edward Teller in the early 1980s, the proponents claiming that such lasers, energized by a nuclear explosion, could destroy all incoming missiles. Since X rays cannot penetrate far into Earth's atmosphere, the system would necessarily be space-based. The Livermore group maintained that a fully-weaponized X-ray laser system could be engineered and delivered in a short time, although some people at Livermore were dubious. After twenty years and an enormous amount of spent money, the claims of the Livermore group about the efficacy and easy engineering of such a system now appear to have been in error and may be another occasion of Tellerian bombast in the corridors of power.

But the question still stands: Is it likely that North Korea or Iran would attempt a suicidal nuclear attack against the United States absent an ultimate threat to their own security? Fanatical terrorism remains a more likely origin of a nuclear attack against the United States or other industrialized countries.

The push by the Bush administration to develop an active antimissile defense has no support in logic. The Bush administration has not made an effort to publicize the relevant analyses by the scientific community, and one suspects the push for missile defense by the Bush administration is being made either because it sounds good politically or because it's an effective mechanism for transferring taxpayer money into the hands of military contractors.

TERRORISM AND NUCLEAR WEAPONS

TERRORISM—NUCLEAR, CHEMICAL, BIOLOGICAL—REMAINS the most serious threat to the United States and other industrialized democracies, and although democracy is based on the idea of an "enlightened" electorate, governments are not in the habit of fostering enlightenment if such enlightenment interferes with short-term political objectives. At the present time, the pronunciamentos of government officials that domestic security against terrorism is in good hands and sufficient can be characterized as at best a twisting of reality. It's a public disservice to tell the public they are safe when science indicates they are not safe. In dangerous times, short-term political expediency can be a lethal strategy.

The realities concerning terrorist use of a crude "hand-delivered" nuclear device are disturbing. Administrative chaos in Russia following the collapse of the Soviet Union has made the enormous nuclear arsenal of the Russians a potential source of nuclear devices and nuclear materials for terrorists with financial resources. Approximately 600 tons of weapons-grade uranium or plutonium are believed to be stored across the former Soviet Union under poor security conditions. The International Atomic Energy Agency has reported that illegal trafficking in nuclear materials has doubled since 1996, and the agency has counted 370 confirmed cases of smuggling of such materials between 1993 and 2001. A U.S. State Department study has noted that as many as 130 terrorist groups worldwide have expressed interest in obtaining nuclear capabilities—among them Osama bin Laden's Al Qaeda group. Some analysts maintain that with a softball-size lump of enriched uranium, a few materials readily available at a Radio Shack store, and a competent university engineering graduate at hand, terrorists would have a reasonable chance of making a crude nuclear weapon.

Former Soviet scientist Vadim Birstein has publicly declared that enriched uranium has already been smuggled out of Russia by the Russian Mafia. If this is true, the most prudent course to protect the United States against a nuclear disaster is to put aside antimissile defense, Star Wars, X-ray lasers blowing up missiles in space, and concentrate security resources on the possible smuggling of nuclear materials or a nuclear device into the United States by ordinary means. Promoting an elaborate and costly missile defense strategy instead of a more rational security program leaves a serious public danger unchecked.

It's unlikely that terrorists would try to do damage with radioactive material not coupled to an explosive. Radioactive material alone is unsuitable for exposing large numbers of people to large doses of radiation, and small doses may not affect health for many years. Explosives packaged with radioactive material are the simplest nuclear device, and such a "dirty bomb" can be devastating to a city. There might not be much physical damage from such an explosion, but the fear and panic about radioactive contamination would be intense, disruptive, and could last for years.

A related threat concerns a terrorist attack on a nuclear power plant. Provided power-plant security could be breached, destructive radioactive fallout could be caused by using only a conventional explosive. In U.S. nuclear power plants, the reactor core is surrounded by thick stainless steel inside a concrete containment building. The most dangerous explosion is the release of an atmospheric plume of radioactive material through a breach in the reactor core. Such a release could have immediate local health effects, and the release of radioactive iodine could produce, at great distances from the power plant, thyroid cancer many years later.

If an actual nuclear device rather than a mere dirty bomb is used, the outcome in a large city will be complete havoc. Once a nuclear device is in the hands of a terrorist group inside a large city, even a bungled explosion can have serious consequences for the population.

The destructive effects of a nuclear blast are derived from the air blast as well as from the heat. An increase in atmospheric pressure from normal pressure at 15 pounds per square inch to 16 pounds per square inch will break glass. At 27 pounds per square inch, the predicted fatality rate among persons close to windows is 50 percent. The nuclear fireball will cause flash and flame burns as well as burns in the subsequent firestorm. Looking at the fireball, even from several miles away, will cause temporary or permanent blindness. Ionizing radiation is released as an intense pulse during the first minute and as fission and activation products after the first minute. A ground-level burst injects a large amount of radioactive soil and other materials into the atmosphere, causing radioactive fallout that may extend over an area of hundreds of square miles. Unless people are protected from exposure in a shelter or are evacuated, the fallout can be lethal at greater ranges than either the blast or the fireball. The devastation in a city like New York or London or Chicago or Bristol from even a bungled ground-level nuclear explosion would clearly be catastrophic.

For sixty years the clock on the cover of the *Bulletin of the Atomic Scientists* has depicted the closeness of civilization to nuclear destruction by the distance between the hands of the clock and midnight. Although a terrorist nuclear device exploded in a major cosmopolitan city will not destroy civilization, such a catastrophe would put Americans and the civilized world in a state of panic whose consequences are unpredictable. As a resident of such a city, I have yet to see any official advice or preparation that would be helpful to ordinary people, no advice, no special medical facilities, no shelters, no evacuation plans—the public remains ignorant. Are the local officials as ignorant as the public, or are they too busy watching the sky for guided missiles? The first terrorist nuclear bomb will be exploded at ground level, in an apartment or small house, in a place like Brooklyn or Islington, and since no one in the neighborhood will survive, the perpetrators will most likely never be identified. Nuclear catastrophe will not come from the sky.

CHEMICAL TERRORISM

CERTAINLY, ONCE GROUPS THAT WILL kill ordinary people for no reason except to produce chaos in civilization exist—groups that will have their own members commit suicide as human bombs—the methods available to such groups are so varied, a terrorist group can pick any method it likes to suit its organization and strategy. But despite the public warnings of some government officials, all methods are not the same, and the probabilities and effects of use vary greatly.

Chemical agents, for example, generally differ from biological agents in several key ways. Chemical agents can cause damage almost immediately and often have no effective therapy, necessitating rapid discovery of an attack (within minutes or preferably seconds) and rapid evacuation. In contrast, biological agents act more slowly (except for pure toxins), and accurate identification of the infectious agent and its exposure range is critical.

Sarin (o-isopropyl methylphosphonofluoridate) is a highly toxic nerve agent chemical that can be fatal within minutes to hours. It was first synthesized in Germany in 1937 as an insecticide, but its potential as a weapon of war was soon recognized by the military. During World War II, Germany prepared thousands of tons of the potent nerve agents tabun and sarin but refrained from using them. Sarin was apparently first used in wartime during the Iran-Iraq conflict in the 1980s. Sarin is a high-potency organophosphate ester. It's a clear colorless liquid, more volatile than other nerve agents, such as soman, tabun, and VX, and it thus presents both a liquid and vapor hazard. In the liquid state, it can rapidly penetrate skin and clothing. As a vapor, it can rapidly penetrate the mucous membranes of the eyes or be inhaled into the lungs to be rapidly absorbed into the body. Sarin exposure causes inhibition of the enzyme acetyl-cholinesterase, and the consequence is massive disruption of the neuromuscu-lar system and rapid death.

In general, chemical attacks can be delivered with almost any spray device, or by an unpredictable method such as that used by the Aum Shinrikyo reli-gious cult that launched two sarin attacks in public places in Japan. The first at-tack occurred in Matsumoto in June 1994. The second attack occurred in a Tokyo subway in March 1995. The terrorists involved in the second attack car-ried diluted sarin solution in plastic bags into subway trains and punctured the bags with sharpened umbrella tips. This released diluted sarin vapor into three lines of the Tokyo subway system. The attack remains the largest disaster caused by nerve gas in peacetime history and illustrates how easily an ill-prepared dis-aster management system can be overwhelmed by chemical terrorism. The at-tack caused twenty-two deaths and affected approximately 5,000 people.

Since the sarin nerve gas attack in Tokyo in 1995, strategies for detecting and responding to chemical warfare agents in semi-enclosed structures such as subways have been a focus of study. In the case of a chemical attack, sarin symptoms are typically prompt, with coughing, choking, distress, and some-times death occurring as quickly as seconds after exposure. Detectors must therefore respond nearly instantly to minimize exposure and guide medical in-tervention. Chemical sensors with quick response times are available commer-cially, but the devices have a tendency to produce high false-alarm rates. The U.S. Department of Homeland Security is supposed to be funding "the devel-opment of strategies and technologies to monitor subways and airports and to rapidly restore critical infrastructure to operational status to alleviate economic impacts."

Although most government officials focus on nerve agents, there are many possibilities for terrorist chemical weapons. In general, the same research that produces potent killers of cancer cells (cytotoxins), for example, can be used as the first phase of a program to produce a variety of chemical weapons. The pharmaceutical industry already has a huge database of toxic compounds, and many of these compounds can serve as the starting points for the development of new chemical weapons. It's unlikely that the research and development of a chemical agent would be a terrorist-group undertaking, but terrorist groups with enough cash can easily buy already developed chemical-weapons stocks stolen or developed in clandestine laboratories.

Terrorism experts often emphasize that a chemical attack by terrorists is not likely to cause much loss of life because volatile chemicals readily disperse into low concentrations and liquid chemicals remain localized. But if the focus of a terrorist group is chaos rather than killing, the Tokyo sarin incident stands as a warning: Chemical terrorism agents are easy to come by, easy to deliver, easy to use in an attack, and a successful attack in the transportation system of a large city, or in an enclosed sports stadium or theater, will produce enough chaos to make the attack useful to the terrorists. Meanwhile, when was the last time you read a posted notice advising the public about procedures in case of an apparent chemical attack? Do you own a gas mask? Since an agent like sarin penetrates clothing and skin, do you think a gas mask will be of any use to you?

BIOLOGICAL TERRORISM

THE MEMBERS OF CONGRESS ARE survivors; they understand the importance of action in human events. When a few envelopes coated with anthrax spores were detected by security devices, the members of Congress ran for the exits as fast as their feet could carry them. Unfortunately, ordinary citizens don't have such security devices; what's worse is that ordinary citizens know so little about microbes, microbial biological weapons seem more exotic than nuclear bombs. A bomb, at least, explodes, and everyone knows what an explosion is, and everyone has seen doomsday photographs of fission and fusion explosions, and even some photographs of the devastation of Nagasaki and Hiroshima, maybe also some gruesome photographs of A-bomb survivors. But what is a bacterial spore? What is anthrax anyway? What is tularemia? Are we really in danger of an attack with germs, or is all of this some plot to keep us on edge? Do the covers of news magazines reflect the popular mood or do the covers make the popular mood?

Imagine, for the sake of this exposition, that you're the leader of an international terrorist group. You have a dozen underground terrorist cells at your disposal in Europe, another dozen in Asia and the Middle East, and a few in the

United States. Your objective is to cause chaos in the industrialized world and do it in a way that will make you seem omnipotent. What you want is a spectacular demonstration of your genius and power. Above all, you need a mission and a method that will not fail, since failure is one certain way to lose new recruits and lose financial support. What mission, what method, and what are the odds of success? If your mission is to bring havoc to a world-class city, say London or New York, your last choice is a biological weapon. In fact, if you're any good as a terrorist leader, a biological weapon would not be considered at all: too risky, too problematic, not sensible when a dirty bomb or a poison gas is much easier to manage and more definite in outcome. Any terrorist attack against a city can fail, but the most likely failure is an attack with a biological weapon.

The anthrax episode in 2001 apparently caused the deaths of some innocent people, produced a great deal of confusion and fear, and then settled into the libraries as another unsolved mystery. This episode was not the work of an international terrorist group; such terrorists are not known to send tacky contaminated letters to individual politicians or newsrooms. But the episode suddenly heightened public awareness of the possibilities of biological terrorism, and it certainly made anthrax a famous microbe.

IS THE ANTHRAX DANGER SERIOUS?

The anthrax bacterium is a bacillus. The term "bacillus" is a common Latin word and literally means "a stick." Bacilli are rod-shaped bacteria in a variety of forms, some short and thick, others long and slender, and the end of the cell is often swollen to accommodate its spore.

Anthrax spores were much in the news during the anthrax episode. Spores are specialized cell structures that allow survival of the organism in extreme environments, for example in the complete absence of water. Not all bacterial species form spores. In general, under conditions of nutritional depletion, each spore-forming bacterium forms a single internal spore, and this spore is liberated when the mother cell undergoes destruction. The anthrax spore is a resting system, highly resistant to drying out (desiccation), highly resistant to heat, chemical agents, ultraviolet light, gamma radiation, and many disinfectants. Spores do not divide, and they have no measurable metabolism; they are inert bits of living matter.

When the spore is returned to favorable nutritional conditions, it activates and germinates a single bacterium. So if, as a terrorist, your interest is a bacterial pathogen to be used as a biological weapon, a pathogenic bacterial species that forms spores is a sensible choice, since spores can be dried and packaged and are not easily damaged in transit. Put the spores in a delivery system and once the spores are in an environment with ordinary humidity and temperature and nutrients, the spores will quickly produce infectious bacteria that begin to replicate themselves.

Anthrax bacteria grow long chains of cells aligned along the long axis of the cell. The typical anthrax cell is a rod 1 micron in diameter and 3 to 4 microns long, with square ends and a spore located in the center of the cell. The virulence of anthrax is caused by the secretion of two toxins, one causing edema, and the other, the lethal toxin, stimulating the massive release by the body of certain protein factors that in high concentrations cause sudden death. Anthrax is not an amateur pathogen: It may be a common soil bacterium, but under the right circumstances, it's a deadly microbe. In the terminal stage of systemic anthrax infection, the number of organisms per milliliter of blood is of the order of 10 million to 100 million—the microbe has completely conquered its host.

In the anthrax attacks of 2001, anthrax spores were delivered in at least five letters to Florida, New York City, and Washington, D.C. Twenty-two confirmed or suspected cases resulted. Five people died, all of them over forty-three years of age. All of the identified letters were mailed from Trenton, NJ. The anthrax spores in all the letters were identified as from a single strain (the Ames strain). The specific source of the anthrax cultures used to create the spore-containing powder remains unknown.

Research on anthrax as a biological weapon apparently began more than eighty years ago, but most national offensive bioweapons programs were terminated following the Biological Weapons Convention of the early 1970s. The American offensive bioweapons program was terminated during the administration of President Nixon. A number of nations, however, continued development programs. Iraq acknowledged to the United Nations in 1995 that it was producing and weaponizing anthrax. The former Soviet Union also had a large anthrax production program. There is apparently some evidence for the existence of offensive biological weapons programs in at least thirteen countries.

But what about terrorism? Would anthrax make a good terrorist weapon? Delivering anthrax spores packaged in artillery shells to an opposing army might be considered by military people as a feasible weapon, but as a terrorist weapon to be used against large cities, anthrax and other microbial agents remain far inferior to dirty bombs and nerve gases. The Aum Shinrikyo cult in Japan, for example, the cult responsible for the sarin nerve gas attacks, apparently attempted to develop anthrax as a weapon, but gave up the attempt and used nerve gas instead.

A NEED FOR PUBLIC INFORMATION

WHEN, AS IS USUAL IN nations, war is policy, it's often difficult to distinguish reasonable strategies from politically motivated strategies that are less reasonable absent political agendas. That the industrialized world is now involved in some kind of deadly struggle with fundamentalist militants seems apparent. What also seems apparent is that one or more of these militant groups are intent on demonstrating their power by whatever means they have at hand, including the public decapitation of innocents, the exploitation of mentally unbalanced individuals willing to be suicide bombers, and the use of extreme instruments and methods appropriate more for killing civilians than for winning military campaigns against an enemy.

It's the instruments and methods that derive most directly from science, and any government that seeks to protect its people from terrorist madness needs to bring any relevant science out of its files and make the science public. If people will be living out their lives under threat of imminent catastrophe, they deserve to be fully informed about details, fully informed and allowed to decide themselves whether their government is adequately protecting them. Too much information about the weapons of terrorism and about the basis of our policies, information already known to the terrorists, is hidden from the people by their governments. Hiding such information is a brand of junk science; it may be politically useful, but it's not the way to run a democracy.

13

GLOBAL WARMING: YES, YOUR BEACH HOUSE MAY BE GONE

[Global warming] is the second-largest hoax ever played on the American people, after the separation of church and state.
—U.S. Senator James M. Inhofe (R-Okl.), 2005

We have not inherited the world from our ancestors, we have borrowed it from our children.
—Native American saying

HUMANS ARE A CREATIVE species with a potentially interesting future, but they do have some maladaptive traits. One is to start killing each other en masse at the drop of a handkerchief. Another is to think a man in a three-piece suit always knows more than a man wearing a lab coat. And a third is to believe the Earth is unchangeable. If serious global warming is a near-future reality, our great grandchildren will see the dangers of all three traits in action: The weather on Earth will undergo a catastrophic change; the men in three-piece suits will pick up their profits and run for high ground and cool air; and nations will go to war because they won't know what else to do.

Predicting doomsday may seem easy because of how often we hear about it, but it's a difficult task and profoundly depressing. People would like to know that at least a few of their great grandchildren will make it into the twenty-second century, and it's not easy to tell them such expectations about their great grandchildren may not be justified. At the present moment optimism is unrealistic, and the reason is that some very smart people who happen to be climatologists are warning us we may be in danger, and people who should be listening to them are instead listening to a group of corporate executives who profit from the status quo and who believe that no matter what happens, if they have their pile, they and their families will survive. I suppose anthropologists have a name for this sort of behavior, but whatever one calls it, the end result may be a catastrophe of a magnitude never seen during all of human history. Will the species survive? I don't know. The human species is not indestructible, and it can be most easily destroyed by the Earth itself.

THE GLOBAL WARMING PROBLEM

THE CURRENT GLOBAL WARMING DEBATE in America is a dizzy carnival of junk science, cretinesque politics, media ballyhoo, and corporate deceit that will only become more raucous and insane as we move forward in the twenty-first century. One has the impression that some politicians dip their heads in crude oil before standing up in Congress to do their ritual lambasting of research on fossil fuels, carbon dioxide, greenhouse gases, global warming, and any and all "alarmists" who warn that the status quo may be pushing us into an abyss.

The important questions are not difficult to identify: Has there been a global temperature increase during the past century? If yes, how much of this temperature increase is caused by man-made emissions into the atmosphere producing a "greenhouse effect"? If the temperature rise exists, will it continue, and if it does continue, what are the expected consequences? These are the important questions—all the other questions are derivative.

The global warming issue is essentially a pollution issue, but on a global scale and with potential costs of one strategy or another dwarfing the costs to any single industry or any single government. The stakes are enormous, the politics frenetic, and the science not as clear as many would like. What is clear, however, are the untruths and twisted science offered to the public by fossil-fuel lobbyists, conservative hacks, and some overly zealous environmentalists on the other side. I think the general strategic question is simple: Should we try to ameliorate a possible danger at tremendous cost, or take our chances and maybe suffer catastrophe that would make the tremendous cost seem trivial?

❁   ❁   ❁

THE DETAILS OF THE strategic game can be outlined as follows:

1. Assuming anthropogenic emissions are an important variable, if we act now to reduce global warming by man-made emissions and our action is not necessary, the cost to various industries and economies may be enormous and wasted.

2. If we act now and the action is indeed necessary and ameliorative, we may have an enormous cost but avoid the even higher cost of monumental global catastrophe.

3. If we do nothing and action is not necessary, we save ourselves the enormous cost to various industries and economies.

4. If we do nothing and serious global warming is a real consequence of man-made emissions, a monumental catastrophe may result.

This strategy game, although easy to describe, is not trivial in consequence, and one can think of instances in history when many people had to make comparable personal decisions about the welfare and safety of themselves and their families in the face of impending but uncertain political danger. Certainly, anyone who has ever had to decide whether to become a refugee or whether to emigrate in mid-life to another country knows what this game is all about.

But I want to avoid the policy questions. I want to focus not on the politics or policymaking but on the junk science in the global warming carnival. What are the myths and what is the evidence?

JUNK-SCIENCE THEMES

THERE ARE CURRENTLY TWO MAIN global warming junk-science themes. The first is that the idea of global warming is an international hoax, that there is no evidence for temperature change, no evidence that carbon dioxide emissions are involved, no evidence of a host of conclusions by climatologists. The second junk-science theme is that there is indeed evidence for temperature change in the past hundred years, but that there is no evidence the change is caused by carbon dioxide acting as a "greenhouse gas," and therefore no reason to reduce emissions. Proponents of the second theme believe the temperature change is merely a natural variation in surface temperature due to one or more natural causes rather than to anthropogenic effects.

One of the proponents of the second theme is Frederick Seitz, which is of interest since Seitz is a scientist (although a solid-state physicist and not a climatologist). Seitz is also well-known to the science community as a former professor at Rockefeller University and a former president of the U.S. National Academy of Sciences. But we should note the academy has seen fit to publish a statement dissociating itself from Seitz's views about global warming.

There are many different individual proponents of these two themes, which are essentially junk-science themes that involve either ignoring, twisting, or sometimes even falsifying data. Senator James M. Inhofe's arguments coincide with the first two and often seem the least rational, the apparent method to formulate every conceivable proposition against the views of climatologists, even if his propositions are inconsistent with each other, maybe with the idea that different people with different critical faculties will be swayed by different arguments and then unite to prevent any action.

But the themes are not worth mentioning unless we know the myths and realities about global warming and its potential dangers.

GLOBAL WARMING MYTHS

MYTH: SCIENTISTS VIGOROUSLY DISAGREE OVER whether human activities are responsible for global warming.

This is a general canard of those opposed to any action, the focus on "disagreement" reminiscent of a similar focus by creationists on supposed disagreements among evolutionary biologists.

Let's consider the question of agreement or disagreement in the context of the organization called the International Panel on Climate Change (IPCC). What is it? Organized under the auspices of the United Nations, the IPCC periodically presents a voluminous report on global climate change to the public. Every five years, approximately one hundred member governments propose the names of their own selected climate scientists, and from the thousands of nominations, the scientist leadership of the IPCC then selects several hundred scientists for each of three working groups, the selection based on the publications of these scientists in scientific journals. Each scientist is assigned responsibility for the entire scientific literature on a particular aspect of the problem. Other scientists are drafted as reviewers and critics, and by the end of the five-year cycle, at least 1,500 experts, including nearly every important climatologist on the planet, is involved in one way or another in the process of creating the final reports.

What does this international group of climatologists say about whether human activities are responsible for global warming? Here are their words from the last assessment report in 2001:

Since the late 1950s (the period of adequate observation from weather balloons), the overall global temperature increases in the lowest eight kilometers of the atmosphere and in surface temperature have been similar at 0.1 degree centigrade per decade.

Furthermore:

The atmospheric concentration of carbon dioxide (CO_2) has increased by 31 percent since 1750. The present CO_2 concentration has not been exceeded during the past 420,000 years and likely not during the past 20 million years. The current rate of increase is unprecedented during at least the past 20,000 years.

Unless one proposes a group of mysterious aliens continually visiting us during the past 250 years to spritz carbon dioxide into our atmosphere, the only sensible conclusion from the above data is that the increase in atmospheric carbon dioxide concentration has been due to anthropogenic emissions, particularly those associated with the Industrial Revolution.

The IPCC climatology group furthermore points out: "About three-quarters of the anthropogenic emissions of CO_2 to the atmosphere during the past 20 years is due to fossil-fuel burning. The rest is predominantly due to land-use change, especially deforestation."

There is hardly any disagreement among the most reputable climatologists in the world that human activities are responsible for the current warming trend. Frederick Seitz, who is not a climatologist, disagrees. He says, "The scientific facts indicate that all the temperature changes observed in the last hundred years were largely natural changes and were not caused by carbon dioxide produced in human activities."

The "facts" alluded to by Seitz concern the absence of a "greenhouse fingerprint," which is simply the heating of the atmosphere that should be caused by the presence of carbon dioxide. Seitz says, "What do the temperature measurements tell us? Satellites and high altitude balloons have been monitoring the temperature of the atmosphere throughout the 1980s and 1990s, and they show no warming trend. The greenhouse 'fingerprint' is missing."

But the contrary is true. Here are the words of IPCC scientists on the warming of the atmosphere:

> *Since the start of the satellite record in 1979, both satellite and weather balloon measurements show that the global average temperature of the lowest 8 kilometers of the atmosphere has changed by +0.05 degrees centigrade per decade.*

The change is small, but it's certainly a change, and the statement by Seitz is not valid. There is indeed already a "greenhouse fingerprint" in Earth's atmosphere. The consensus of the world's leading climatologists, 1,500 scientists, is that human activities are responsible for global warming.

And the IPCC climatologists are not alone in this conclusion. The U.S. National Academy of Sciences report on the subject states: "Greenhouse gases are accumulating in Earth's atmosphere as a result of human activities, causing surface air temperatures and subsurface ocean temperatures to rise." Similarly, the American Meteorological Society, the American Geophysical Union, and the American Association for the Advancement of Science all have issued recent statements concluding that the evidence for human modification of climate is compelling.

MYTH: CARBON DIOXIDE IS not a pollutant. It's necessary for life.

So it is. Carbon dioxide is certainly necessary for life. Without carbon dioxide in the air, most plants die. Without most plants, most animals die. Without most animals, you and I die. This is a favorite argument of U.S. Senator Inhofe, who thinks the idea of global warming is a hoax perpetrated on the American people. Since Senator Inhofe is well aware (he says he has spent "years" analyzing the problem of global warming), he must know that although carbon dioxide is necessary for life, it's also a gas that in the atmosphere prevents the loss of infrared heat, which produces the so-called greenhouse effect, which means that in the atmosphere too much carbon dioxide can produce a disaster.

The term "greenhouse effect" refers to the blockade of longwave (infrared) radiation from the Earth into space by trace constituents of the atmosphere, primarily water vapor, carbon dioxide, ozone, methane, nitrous oxide, and various chlorofluorocarbons—gases referred to collectively as the "greenhouse gases."

What happens in global greenhouse warming is similar to what happens in an actual greenhouse: The glass of the greenhouse allows sunlight to enter, the sunlight warms the interior of the greenhouse, but the glass blocks radiant heat (infrared radiation) from escaping, and the result is the interior of the greenhouse is warm in the middle of winter.

The sensitivity of the biosphere to the greenhouse effect has existed throughout the history of life on Earth: The infrared radiation-blocking gases return radiated heat to the ground, accounting for approximately 70 percent of the net input of energy to the Earth's surface.

Carbon dioxide is the major greenhouse gas on the planet Venus. On that planet, a high atmospheric concentration of carbon dioxide apparently produced a runaway greenhouse effect that raised the surface temperature of that planet to 880 degrees Fahrenheit. No one believes such a calamity will happen on Earth, not in the near future, but to tell the American people that carbon dioxide is not a pollutant and therefore carbon dioxide emissions are not of concern for climate change is a gross distortion of reality.

MYTH: THE BALANCE OF the evidence offers strong proof that natural variability is the overwhelming factor influencing climate.

Reality: There is absolutely no "strong proof" that natural variability is the "overwhelming factor" influencing climate in the recent past, present, or near future, and that is the time frame at issue, and not time frames of thousands or millions of years. Are we to assume that proponents of this myth are deliberately playing with ambiguities in order to pursue a personal agenda?

What is the evidence concerning so-called natural climate variability? How sensitive is the climate to changes in solar irradiance, atmospheric aerosols, greenhouse gases, and other climate forcings? To answer this question, researchers must first know the true extent of past climate fluctuations. The

changing temperatures over past centuries and millennia have been reconstructed by regressing annually resolved climate proxy records—for example,
from ice cores and tree rings—against recent thermometer measurements.

Environmental change involves jumps, fluctuations, and trends, the environment changing through the operation of the internal machinery of the ecos-
phere (biosphere), and through the external agencies of cosmic and geological
forces. Evidence of past environmental change, almost always incomplete, derives from geochemical, physical, biological, historical, and instrumental
sources. In recent years, high-speed computers have allowed researchers to ma-
nipulate complicated and reasonably realistic models of environmental
change, with modeling particularly useful for studying changes in sedimentary
basins, biogeochemical cycles, and climate. General circulation models, run
with appropriate boundary conditions, predict climates of the past, and these
predicted climates can be compared with paleoclimatic indicators.

It has long been known that on timescales of tens of millions of years, intervals of continental glaciation were interspersed with intervals of little or no ice.
The magnitude of warmth during these warm intervals is impressive: For example, at times during the Cretaceous period (approximately 65 to 146 million
years ago) duck-billed dinosaurs roamed the northern slope of Alaska, and deep
and bottom waters of the ocean, now near freezing, could reach a balmy 15 degrees Celsius.

Research indicates that the last few million years have been generally cold
and icy compared with the previous hundred million years, but the climate has
apparently alternated between warmer and colder conditions. These alterna-
tions have been linked to changes over tens of thousands of years in the sea-
sonal and latitudinal distribution of sunlight on Earth caused by features of
Earth's orbit. Globally simultaneous (synchronous) climate change, despite
some hemispheric nonsimultaneous occurrence (asynchrony) of the forcing, is
explained at least in part by reduced atmospheric carbon dioxide during colder
times in response to changes in ocean chemistry. We currently live in one of
the warmer times of these orbital cycles. In the past, the coolest times brought
glaciation to nearly one-third of the modern land area.

But natural climate forcings are not the issue in the current global warming
problem, and there is no evidence that such forcings are the explanation for the
changes that have occurred in the recent past and that will apparently occur in
the near future. The fossil-fuel industry is opposed to regulation of fossil-fuel
emissions, and therefore opposed to any science that leads to a conclusion that

regulation may be necessary. Manipulation of the facts of climatology to suit an opposition to any change in governmental climate policy, particularly any change that would restrict fossil-fuel emissions, is junk science in the service of an industrial agenda.

MYTH: NO ONE HAS seriously demonstrated any scientific proof that increased global temperatures would lead to the catastrophes predicted by "alarmists."

Reality: This argument is specious, since no one can provide "proof" that any event will occur in the future. All that can be provided are statements about estimated possibilities and probabilities, with viewpoints substantiated (or not) by past climate history and models that simulate climate dynamics. At the present time, the IPCC is projecting that the globally averaged surface temperature will increase (depending on the model used) by 1.4 to 5.8 degrees centigrade over the period 1990 to 2100. This projected rate of warming is much larger than the observed changes during the twentieth century and is "very likely to be without precedent during at least the last 10,000 years."

What are the expected effects of a global temperature increase in the range 1.4 to 5.8 degrees centigrade? The IPCC summarizes potential adverse effects as follows:

- A general reduction in potential crop yields in most tropical and subtropical regions for most projected increases in temperature.
- A general reduction, with some variation, in potential crop yields in most regions in mid-latitudes for increases in annual average temperature of more than a few degrees centigrade.
- Decreased water availability for populations in many water-scarce regions, particularly in the subtropics.
- An increase in the number of people exposed to vector-borne (e.g., malaria) and water-borne diseases (e.g., cholera), and an increase in heat-stress mortality.
- A widespread increase in the risk of flooding for many human settlements (tens of millions of inhabitants) from both increased heavy precipitation events and sea-level rise.
- Increased energy demand for space cooling due to high summer temperatures.

These are the projected consequences of moderate climate change during the twenty-first century, but climatologists also believe the potential exists for "future large-scale and possibly irreversible changes in Earth systems resulting in impacts at continental and global scales." These possibilities depend on various climate-change scenarios and include significant slowing of the ocean circulation that transports warm water to the North Atlantic, large reductions in the Greenland and West Antarctic Ice Sheets, accelerated global warming due to carbon cycle feedbacks in the terrestrial biosphere, and release of terrestrial carbon from permafrost regions and methane from hydrates in coastal sediments.

Climatologists believe that if these changes were to occur, their impacts would be widespread and sustained. For example, significant slowing of the oceanic thermohaline (heat plus salinity) circulation would impact deep-water oxygen levels and carbon uptake by oceans and marine ecosystems and would reduce warming over parts of Europe. Disintegration of the West Antarctic Ice Sheet or melting of the Greenland Ice Sheet could raise global sea level up to three meters each over the next thousand years, submerge many islands, and inundate extensive coastal areas. Depending on the rate of ice loss, the rate and magnitude of sea-level rise could greatly exceed the adaptation capacity of human and natural systems. Release of terrestrial carbon from permafrost regions and methane from hydrates in coastal sediments, induced by warming, would further increase greenhouse gas concentrations in the atmosphere and amplify climate change.

Are these the rantings of "alarmists" or the concerns of several thousand trained scientists, experts in oceanography, atmosphere science, and climatology? Given the size and consensus of the scientific community involved, is it too much to ask that anyone opposed to these consensus views present the public hard science in standard journals rather than ambiguous mumblings about "lack of agreement among researchers" and "absence of proof concerning harmful effects"? Are we to base public policy concerning a situation that may compromise the future of millions on the ignorance of politicians and the special interests of a few industries? What is the most prudent course in the public interest?

MYTHS ABOUT CLIMATE MODELS

THE ULTIMATE CANARD IS THE claim that the IPCC predictions are baseless be-cause "climate models are highly imperfect instruments." Yes, indeed, the pre-dictions are as imperfect as the war games and defense scenarios played out in the Pentagon by Defense Department specialists. In the presence of complex-ity and in the absence of scientific certainty about variables, one uses models to compute possible outcomes, and since every computational model involves assumptions concerning unknowables, models are imperfect predictors, imper-fect in economics, imperfect in defense strategy, imperfect in corporate finan-cial planning, imperfect in clinical medicine, and imperfect in climatology.

Climate models range from simple zero-dimensional models to elaborate three-dimensional models, but in general there are four main types of climate models.

Energy-balance models predict the change in surface temperature resulting from a change in heating, with the constraint that the net flux of energy remains unchanged.

Radiative-convective models compute the vertical (one-dimensional) temper-ature profile of a single column of air. The predicted temperatures are usually global averages.

Statistical-dynamical models usually deal with a two-dimensional slice of air along a line, commonly a line of latitude. Such models thus combine the latitudinal dimension of energy-balance models with the vertical dimension of radiative-convective models.

General-circulation models are the most sophisticated climate models cur-rently in use. These models consider three-dimensional blocks of air as the blocks move horizontally and vertically through the atmosphere. There are three types of general-circulation models: One type predicts the state of the atmosphere; a second type predicts the state of the oceans and ranges from simple mixed-ocean models to full ocean general-circulation models; and, fi-nally, coupled atmosphere-general-circulation models consider the atmosphere and oceans simultaneously.

If the canard about "imperfect instruments" is to be taken seriously, then policymaking needs to be suspended in every domain, and not merely in those domains that may impact the fossil-fuel industries. Yes, climate models are im-perfect instruments, but if the planet is headed for disaster in the twenty-first century, the warnings provided by our imperfect instruments are all we have, and it's in our interest to take them seriously and not dismiss them with a wave of the hand as "imperfect."

RECENT HISTORY OF CLIMATOLOGY

AN UNDERSTANDING OF THE RECENT history of climatology is helpful in evaluating what has been happening in this field. In 1896, the Swedish physical chemist Svante Arrhenius (1859–1927) published the first calculations of global warming from human emissions of carbon dioxide. Not much attention was paid to his calculations as two world wars and rapid industrialization occupied the attention of the major nations.

In the 1950s, climatologists began using primitive climate models to predict climate, and the idea of global warming due to carbon dioxide emissions came alive again. But the models were primitive, and computing power too limited to use more sophisticated models for predictions.

Research continued, and by the 1970s, computer models were showing a temperature rise of several degrees for doubled atmospheric carbon dioxide concentrations.

In the 1980s, the first measurements of an actual global warming trend began to appear.

In 1988, the Intergovernmental Panel on Climate Change (IPCC) was established and soon issued the first of three sets of reports.

In 1990, the IPCC concluded that Earth had indeed been warming, and that much of the warming might be due to natural processes.

The second set of IPCC reports, in 1995, changed emphasis and concluded that the balance of evidence suggested a discernible human influence on global climate.

The most recent IPCC report, in 2001, had nearly every climatologist in the world agreeing that the world was rapidly getting warmer, and that strong new evidence indicated that most of the observed warming over the past fifty years was probably due to an increase in greenhouse gas concentrations.

This chronology is important in understanding the validity and impact of the criticisms that have been leveled against the idea of global warming. Fearful of any regulation, the relevant industries have always opposed the idea of global warming, and their opposition was supported by the media and by a number of scientists through the 1970s, 1980s, and 1990s, and also supported by the American government. Everything changed after the 2001 IPCC reports, when new evidence and new conclusions silenced most of the critics, even those funded by industry. These days, not much is heard from Frederick Seitz or S. Fred Singer, another critic, and rather than scoffing at the idea of global warming, the media, particularly Hollywood, have been playing it up with a consequent increase in public interest. Worldwide, the only major holdout is now the Bush administration and certain conservative people in the U.S. Congress. They continue to play down global warming, continue to serve the interests of industry, and continue to refuse any serious international cooperation to curb fossil-fuel emissions.

At the present time, in its anti–global warming stance, a stance based on twisted science, America is alone—and also the largest producer of greenhouse gas emissions. In a hundred years, what story will be told about America and global warming? We're the richest country on Earth, a dynamo of human invention, energy, and commerce. Is it our intent to keep filling our pockets even if both we and the world perish in the process?

PART FIVE

RELIGION, EMBRYOS, AND CLONING

14

CREATIONISM: THE WORLD AS AN EGG

I believe that God created Heaven and Earth. . . . I don't believe it's demonstrably true that we have descended from
apes. I don't believe it. I do not believe all that.
—PAT BUCHANAN, U.S. PRESIDENTIAL CANDIDATE

The creationist, whether a naïve Bible-thumper or an educated bishop, simply postulates an already existing being of
prodigious intelligence and complexity.
—RICHARD DAWKINS, BIOLOGIST

SOMEWHERE IN THE HEARTLAND, a ten-year-old boy dreams. He lies on his back on the grass and he gazes at white clouds in a blue sky. How fast do the clouds move? Is the Earth really like a ball? What makes the Earth turn? Why is the sky blue and not yellow? Where do human beings come from? What are germs? A few months ago, he begged his father to buy him a toy microscope. He looked at pond water and saw some moving specks and wriggling things. Were they germs? After a few days, he took apart the microscope and tried to make a telescope to look at the Moon. When his father learned of it, he was angry. His parents said he was ungrateful. He likes to walk along the road near his house and keep his eyes and ears open for snakes. Snakes move very fast in the grass and catching a snake is a challenge. Why is the grass green?

What should we do with this boy in school, or with his sister, who asks as many questions as he does? Should we teach them myths and lies about the world because the myths and lies give us comfort?

DARWINISM IN AMERICA

DARWIN WAS AWARE THERE WOULD be trouble, but it's doubtful he thought the trouble in America would be endless. Alexis de Tocqueville, on the other hand, writing a generation before *The Origin of Species*, knew a few things about the American people. "In America," de Tocqueville wrote in 1835, "religion is perhaps less powerful than it has been at certain periods in the history of certain peoples, but its influence is more lasting."

More lasting, indeed. Nearly 150 years after *The Origin of Species*, the conflict between science and the Bible continues to fester in America, while in Europe, the ancestral home of American religion, the scientific explanations of the history of Earth and of the evolution of life upon the planet have been generally accepted without furor, at least by the general populace. In 1996, Pope John Paul II himself, addressing the Pontifical Academy of Sciences, finally acknowledged, "the theory of evolution is more than a hypothesis."

An advance, perhaps, but dogma must be preserved: In the same address the Pope cautions that the idea that the mind emerges from living matter or is an epiphenomenon of matter is "incompatible with the truth about man." And then in 2005 Christoph Schoenborn, the Roman Catholic cardinal archbishop of Vienna and the lead editor of the official 1992 catechism of the Catholic Church declared to the world that "evolution in the neo-Darwinian sense—an unguided unplanned process of random variation and natural selection—is not [true]," the cardinal stating: "Any system of thought that denies or seeks to explain away the overwhelming evidence for design in biology is ideology, not science." This is certainly a strong statement against all of modern science, with twisted logic that calls the Church scientific and scientists ideologues.

So if ordinary Europeans are generally more accepting of neo-Darwinian evolution than ordinary Americans, any marriage between the Church and science is not yet in the works—not even an announced engagement. Man's body may have evolved, but his soul is still held fast by the ecclesiastics. And neo-Darwinism, the merging of classical Darwinian evolution by natural selection with modern genetics, the cornerstone of modern biology, is summarily rejected. The Church apparently accepts evolution and the descent of man from animals but declares the process is all by design of a Creator rather than by natural selection of random variation. (The archbishop's denunciation of neo-Darwinism suggests he's unaware that the idea of natural selection of random variation is classical Darwinism.)

In contrast to the focus of the archbishop of Vienna, the fundamentalist religious zealots in America (who take the Bible literally) remain uninterested in disputes about the mechanism of evolution, which is rejected in toto. Not content to enjoy their extreme religious beliefs about creation and the origin of man in private, they want to impose their beliefs on others, force the teaching of those beliefs in the public schools, force a corruption of science in the schools that will handicap any child who seeks to become a medical doctor or a molecular biologist or just a person with even a modicum of understanding of the real world.

One of the silliest canards is that evolution is "only a theory," a speculation no better than other speculations. The charge can come only from someone with little understanding of general science or of the science of biology. A theory in science is not a guess. Evolutionary theory is no more a guess than atomic theory. We have so much evidence supporting not only the idea of the evolution of life forms but of the process of natural selection, that only a person with a closed mind can possibly ignore the obvious conclusions about the validity of the theory. Of course, many details are still debated, but the general idea of Darwinian evolution of life on Earth is as universally accepted in science as the idea that a water molecule consists of an oxygen atom and two hydrogen atoms.

So why is evolution still called a "theory"? The answer is simple: The word "theory" in science has a special meaning. It's a strong conclusion, rather than a direct observation. You don't need a theory that proposes that the sun exists, because you can see the sun with your own eyes. But if we want to know what's happening inside the sun, we need a theory for that because we can't see it and we can't go there. Anyone who wants to restrict their ideas and their life to things that are not theories had better turn off all the lights and electricity and heat in their house, because all of it is based on "theory." And they also need to avoid doctors and hospitals, because nearly all of medicine is also based on "theory." Satellites circle the Earth based on theory, and your cell phone is an application of half a dozen theories in physics. Sure, evolution is a theory, but it's one of the strongest and most widely accepted theories in modern science.

Canards about evolution are common among creationists. Creationism and its offshoot, intelligent design, are not science, they are nonscience, the work of science nullified, and as such they are of no real consequence to science, since they offer nothing that can be useful in the exploration of the real world. Creationism and intelligent design are of consequence to politics, not to science—they are political problems and not scientific problems. The public must finally decide whether it wants its children to be taught real science or junk science, and so far it looks like a decision for real science will not be made without enormous pressure from the American intellectual community.

CREATIONISM IS A POLITICAL PROBLEM

IN CONTRAST TO WHAT MANY believe, it's doubtful that science has much of a role to play in this struggle, since it's not a question of debating fundamentalists or converting fundamentalists or convincing fundamentalists—their position is based on received belief and not on reason, and their position is therefore invincible to reason. The problem of creationist and intelligent-design junk science in the schools remains a political problem that requires a political solution.

Some people may dispute this view and argue that it's necessary for scientists to debunk creationism and intelligent design, as if such debunking has never been accomplished. The scientific community in fact now finds such debunking as trivial as it would the debunking of a view that tomorrow the sun will turn green and assume the shape of a cube. The Earth is approximately four and a half billion years old, all life on Earth evolved from simpler forms, and humans evolved from other animals. The truths about the world are more important than the fantasies of any school board in any town anywhere.

The question remains: What do we teach our children? In a country of nearly 300 million people, there will always be some people who, because of religious conviction, believe the Earth is as flat as a pancake, a few thousand years old, and resting on the backs of four giant elephants. Their belief is unfortunate. What is even more unfortunate is teaching our children that such beliefs, because they are religious, deserve respect. Are the schools to be bazaars of babble, where myths and delusions with or without religious vintage are taught to children as "science" alongside real science?

THE SEVERAL FORMS OF CREATIONISM

THERE ARE SEVERAL FORMS OF creationism, and the differences between them are significant.

Fundamentalist creationism ("young Earth" creationism) takes the Bible as literal history and proposes the Earth is only 6,000 years old. Since the Bible is taken literally, evolution of life on Earth is rejected, all species considered to be created almost at once. Humans have not evolved but were created de novo, and so on.

Non-fundamentalist creationism is more complicated because of its variations. One form, which I call "higher creationism," accepts all the findings of science in cosmology, biology, etc., but proposes a divine spirit behind it all. So it's accepted that life has evolved on Earth, that man has evolved from animals, but everything in the universe has been created and guided by a Creator in some unspecified and indirect way.

Another form of non-fundamentalist creationism is so-called intelligent design theory. This theory proposes an old Earth and an old universe, but a more direct involvement of a Creator, an "intelligence" of some kind to explain the appearance of complex design in the natural world, and particularly in biological systems. This form of non-fundamentalist creationism essentially rejects the idea of evolution by natural selection of random variation. As discussed previously, the Catholic Church is apparently now favoring intelligent design theory, although until now it was believed to be supporting what I've called "higher creationism."

Now let's discuss these various forms of creationism.

THE DANGERS OF "YOUNG EARTH" CREATIONISM

"YOUNG EARTH" CREATIONISM IS NONSCIENCE AND to accept its views in the laboratory would rip apart the entire fabric of modern science, biology and chemistry and physics, and would return us to the time at which the Book of Genesis was written, which means a loss of 6,000 years of human intellectual history. I find the loss unacceptable and I'm not willing to make the trip backwards in time.

Young-Earth creationists believe that evolution of animals and man from animals is a false idea. The prime difficulty for this belief is that many aspects of the real world, including all of medicine, are based on the idea of evolution.

I use the phrase "real world" a great deal in this book. What is it? One dramatic presentation of the real world is the intensive care unit of a modern hospital, the large room with its tubes, wires, machines, monitors, beds, patients, and watchful nurses, the entire assemblage focused on the task of keeping the people in the beds alive, all of it the application of modern science to preserving human life. One challenge to any creed that seeks to twist science to its own ends is to ask what that twisting and the creed behind it have to offer to the intensive care unit. When a creationist has a child who is suddenly ill, the child (one hopes) is rushed to a hospital to be given every advantage that modern medicine can offer. All the knowledge, all the instruments, all the drugs would be vaporized in the absence of basic biological science, the science that sits on evolution as its guiding principle. Do creationists realize what they are asking for when they seek to destroy the bedrock of modern biology and medicine?

What many young-Earth creationists apparently do not understand is that biologists can be "wrong" about evolution only if the instruments and methods they use are deceiving them. And since those instruments and methods are applications of fundamental physics and chemistry, refusing the measurements of those instruments and methods essentially refuses all of physics and chemistry—all of it, from the physics of the cosmos to the chemistry of a glass of water—all our understanding of the natural world wiped out, and all the applied sciences based on that understanding wiped out as well. So not only is medicine removed, but all of technology as well.

The issue is all of science teaching and not just of biology teaching. Science may be split into categories for the convenience of teachers and administrators, but in truth it's one undertaking, one endeavor, one enterprise. Those who seek to destroy science for religious reasons need to have their knowledge of consequences challenged. They need to be made to understand that without science their own temples may collapse on their heads, since it's science and applied science that allowed construction of the temples in the first place.

THE EVIDENCE AGAINST A YOUNG EARTH

THE DEMONSTRATION THAT YOUNG-EARTH CREATIONISM is junk science is most easily made by considering the age of the Earth as provided by scientific evidence. The most accurate way to find the age of the Earth is by bringing a sample of it into a laboratory and determining its age by radiometric dating—by measuring the amounts of various radioactive elements remaining in the sample. The oldest Earth rocks so far discovered and dated are approximately 3.9 billion years old. This does not indicate that the Earth formed 3.9 billion years ago, since the Earth is an active planet with a crust continually destroyed and reformed. The age of the oldest rocks reveal only that the Earth is at least 3.9

billion years old. In contrast, the rocks brought back by the Apollo moon land-ings have proved to be 4.48 billion years old, which means our solar system must be at least 4.48 billion years old. Meteorites that land on Earth are a still further source of material, and the oldest of these is approximately 4.6 billion years old, which is widely accepted by physicists as the age of the solar system. The general consensus in astrophysics is that our sun is approximately 5 billion years old, and that all the bodies of the solar system formed at approximately the same time about 4.6 billion years ago.

It's most important to understand that determination of the age of the Earth, the moon, and meteorites is a problem in physics and chemistry, a problem whose solution required methods not available until the second half of the twentieth century. Young-Earth creationism, which asserts that the Earth is only 6,000 years old, is at war not only against biology, but also against physics and chemistry, which makes the war a war against all of science. People who speak of the "Darwin war," as if the war is a war only against biology, are missing a point: This is a war against science, essentially a religious war against reason.

Since young-Earth creationists are fundamentalist creationists, they take the Bible as a science text, assert that the Earth is only 6,000 years old, that life on Earth is only 6,000 years old, that all species were created at once, that dur-ing the flood all species except one male-female pair from each were destroyed, and so on. As I have pointed out, to accept any of this is to deny the validity of all of science because all of science is a single evidentiary fabric of interde-pendent observations. If, when the Bible was written, the people who wrote it had described the Earth to be in the shape of a chicken egg, present-day cre-ationists would no doubt assert that to be the case, no matter all the evidence, instruments, and so on, including the eyes of the astronauts on the moon, and the cameras on the moon, that demonstrate a spherical rather than an egg-shaped Earth. As an intellectual effort, fundamentalist creationism can be dis-missed as myth in explicit disagreement with fact, and that's the end of it.

HIGHER CREATIONISM CONTRADICTS SCIENCE

NON-FUNDAMENTALIST HIGHER CREATIONISM IS ANOTHER matter. In this form of creationism, there is no literal reading of biblical texts, no "young Earth" fun-damentalism, and the problem of conflict between religious assertions and modern science is largely avoided. Instead, the personal comfort provided by a belief in a Creator is retained by proposing that all the facts concerning the history of the Earth and the history of the universe are acceptable, but the entire cosmos, including man within it, is itself the work of a Creator, a God, and that God imbues the soul of man. As J. F. Haught, a higher creationist, says: "Acknowledging the reality of the power of the divine would not rule out an explanatory pluralism that gives ample room to Darwinian accounts alongside others." In other words, Haught will accept all of science, including evolutionism, with the proviso that a divine Creator is behind it all.

Higher creationism might seem palatable, but the problem is that Haught's proviso itself negates science. There is no "reality of the power of the divine," no demonstrable reality of it, although among many people there is certainly a wish for the existence of the power of the divine. A central motif in the fabric of science is that one does not introduce supernatural powers or supernatural forces or supernatural entities to explain *anything*. Not knowing the answer to a natural riddle, including the riddle of the cosmos, is not a basis for introducing the supernatural, at least not in real science. Higher creationists may be comforted by their belief in God, divine power, the existence of the human "soul," the presence of God in the human soul, and so on, but their belief is outside the realm of science and is not necessary for the work of science.

To reiterate, because the principle here is so important: Science makes no judgment for or against the existence of a divine power or God or a Creator; the only judgment made by science is that to explain the world as encountered by science, God or some other supernatural entity has not yet been found to be necessary.

Particularly not necessary for the work of science is any "ultimate reality" such as that proposed by the higher creationist J. F. Haught, who declares: "It is essential to religious experience, after all, that ultimate reality be beyond our grasp. If we could grasp it, it would not be ultimate."

Nothing to reject here, since science cannot deal with anything beyond its grasp. But a caveat is necessary: What is beyond the grasp of science today may not be beyond the grasp of science tomorrow, as many theologians through the centuries have discovered to their dismay.

But as long as higher creationism does not twist science or the teaching of science, there is no political issue of any consequence. As for philosophical issues, the debate is endless and I have no contribution to offer to it except to suggest the usefulness of a "philosophy of ignorance." Some rationalists need to be reminded that premises are often based on what is currently known, and what is yet to be known is more than can be imagined.

As MENTIONED, A SPECIAL form of non-fundamentalist creationism is that proposed by the Catholic Church, although these days the Church's position is not at all clear. The Church has always disdained using the Bible as a literal history of the natural world, so it's position has always been non-fundamentalist. But the Church has also always had trouble with the idea of man as an evolved animal, first rejecting Darwin's ideas, then appearing to accept Darwin's ideas, and most recently apparently rejecting Darwin again. No doubt a clear statement will eventually be issued by the Vatican, but since it took the Church two hundred years to accept Galileo's ideas, it might take the Church as long to accept Darwin's ideas. Meanwhile, the Catholic Church remains a special case: non-fundamentalist, proposing biological design by a Creator, but not accepting Darwinian evolution of the human species. One has the feeling of an enormous confusion, but since the knowledge of the real world gained by modern science is shaking the roots of long-established religion, maybe the confusion is to be expected.

Meanwhile, the Catholic Church is apparently supporting intelligent design theory, discussed below, but without any theological details yet available. Some intelligent design theorists propose that an alien "intelligence," rather than a god might be involved in intelligent design, but it's doubtful the Catholic Church would accept any such proposal.

COMPLEXITY AND INTELLIGENT DESIGN

FIFTY OR SIXTY YEARS AGO, a biologist looking at a large living biological cell through a light microscope could see motions on the surface and in the interior of the cell, motions aplenty and all of it mysterious. It was not until the 1960s that the microscale structures involved in cell movements were roughly identified, not until the 1970s that the biochemistry of these structures was characterized, and not until the 1990s that a clearer picture of the possible intricate movements of the "molecular motors" (motor proteins) of living cells became apparent.

An engineer viewing some of the current models of biological molecular motors will find nanoscale devices involving only a handful of macromolecules, with each device engaged in a precise sequence of repetitive movements — rotations, vibrations, translocations along tracks, linear contractions, etc. — the energy for these motions derived from enzyme-catalyzed reactions, and all of these devices assembled with apparent great precision by synthetic processes controlled by information stored in the genome of the cell. It's quite understandable if the engineer, while looking at a model of the macromolecular assembly evidently responsible for the rotation of a flagellum, the whiplike structure involved in bacterial movement, is flabbergasted. We have apparently crossed a threshold into a world of nanoscale "machinery" in biological cells, and cell biology in the twenty-first century promises to be a source of extraordinary revelations. It's now recognized that the interiors of biological cells are structurally complex, and that this structure is dynamic.

"Motor proteins" are mechanico-chemical enzymes involved in locomotion or transport, and there are several families of such proteins. In general, as mechanico-chemical enzymes, motor proteins convert energy from hydrolysis of nucleotides to mechanical force, and since they are involved in many important cellular events, the molecular details are currently the focus of intensive research.

The bacterial flagellum is an example of motor proteins at work to produce a spectacular living device. Bacteria achieve motion using flagella, which are usually many times the length of the bacterial cell, but so narrow in diameter that they need to be coated with a dye in order to be seen with the light microscope.

At the base of each flagellum is a molecular motor, and bacteria swim by the flagellum rotating like a screw propeller at hundreds of revolutions per second. From crystallographic analysis, several models of the flagellar molecular motor have been proposed, all of them intricate examples of how chemical action can produce organized nanoscale motion.

In recent years, the bacterial flagellum has acquired public fame as a "case-in-point" in creationist attempts to argue for the existence of a Creator. What is at issue here is a non-fundamentalist creationism involving what is called "intelligent design theory."

This idea focuses on complexity of design in living systems and holds that the apparent complexity of "design" in many biological systems is prima facie evidence of "intelligence" in design, and therefore evidence of a Creator. The central concept is that of "irreducible design complexity," a design such that removal of even a single part causes the device not to function. Such devices are said to be incapable of evolving and are thus proof of the existence of intelligent design. The flagellum of the lowly bacterium is considered by these creationists as an example of a living structure whose complexity defies an evolutionary explanation and is a structure that could only have been designed by "intelligence"—in other words, a Creator.

One problem for intelligent design theory is that neither the flagellum nor any other biological structure has been unequivocally demonstrated to be irreducibly complex, which leaves the theory without a real biological structure as a focus. Most biological systems, in fact, especially at the level of molecular genetics, have an enormous degree of evolved redundancy built into the subsystems, structures, and biochemical pathways, a redundancy that makes any truly irreducible design unlikely. From a scientific standpoint, no known biological structure defies evolution in the sense that the structure cannot possibly have evolved.

Does intelligent design theory have any scientific merit? The answer is it does not. The theory is based essentially on the "enigma fallacy," the idea that if one cannot imagine a natural explanation for a seemingly enigmatic event or observation, one must propose a supernatural cause, an "intelligent designer" or Creator or the Wizard of Oz, or some other entity on the road leading straight back to shamanism and witchcraft.

Intelligent design theorists maintain that evolutionism cannot explain irreducible complexity. Their error is not to realize that evolution can proceed quite nicely by accretion of new parts rather than by substitution of old parts, and once the new parts have proved adaptive, the old parts can eventually fade away. The construction of a mousetrap or a watch certainly requires intelligent design, but there is no evidence that the construction of any living system requires intelligent design, and the history of modern science is more or less a continuing demonstration of this.

Intelligent design theorists are not in perfect agreement about whether *any* design requires an intelligent Creator, or whether only irreducible design requires an intelligent Creator, but we will leave that dispute alone.

INTELLIGENT DESIGN IS A REBIRTH OF VITALISM

MICHAEL J. BEHE, A LEADING proponent of the idea of intelligent design, has publicly presented the following as the major claims of the theory:

1. The theory of intelligent design, according to Behe, is not a religiously based idea. Intelligent design says nothing about the religious concept of a Creator, although devout people opposed to the teaching of evolution cite it in their arguments.

2. The physical marks of design, says Behe, are visible in all aspects of biology. Science has shown that the cell, the very foundation of life, is run by machines made of molecules.

3. We have no good explanation, says Behe, for the foundation of life that does not involve intelligence.

4. In the absence of any convincing non-design explanation, says Behe, we are justified in thinking that real intelligent design was involved in life.

Behe closes his essay with the following remarkable statement: "Whatever special restrictions scientists adopt for themselves don't bind the public, which polls show, overwhelmingly and sensibly, thinks that life was designed."

These claims remind one of similar claims made to support "vitalism" at the end of the nineteenth century and well into the twentieth century, before biochemistry and cell biology began their major development. In general, vitalism claimed that the nature of life resulted from a vital force peculiar to living organisms, a force different from all other forces found outside living systems, and this force was proposed to control form and development and to direct the activities of organisms.

There were actually several vitalisms, but they were all opposed to what were considered the extremes of scientific materialism. Often inspired by religion, vitalists refused to accept the mechanistic ideas of Darwin. Underlying most vitalisms was the belief that life was more than mere complex chemistry, since to accept that idea would subject human activity to deterministic explanations. Ignorance at the time of biochemical principles commonly understood today made such phenomena as growth and development mystifying, and causal agencies such as "entelechies" and "vital forces" were invoked to explain what appeared to be baffling phenomena.

The general argument of vitalists was that since physics and chemistry had failed to explain the apparent mysteries of living systems, attempts using only physics and chemistry to explain the differences between animate and inanimate matter were doomed to failure. Contrast this with the attitude of modern biology, as succinctly expressed by Nobel laureate molecular biologist Jacques Lucien Monod (1910–1976): "There are living systems, there is no 'living matter.'"

From a historical-philosophical perspective, intelligent design theory is a continuation of nineteenth-century vitalism dressed up in modern clothes. In the twentieth century, biologists learned that vitalism is not necessary to explain biological phenomena, and now, confronted by new "mysteries" revealed by molecular biology, the junk-science stage play has evidently opened again in various theatres.

WHY INTELLIGENT DESIGN FAILS

THE TWO MAIN DEFICIENCIES OF intelligent design theorists are a lack of appreciation of the minuteness of steps usually involved in evolutionary change, and secondly, a lack of understanding that within any species there is not a single "species genome" but as many different genomes as there are individuals in a population of that species, with each genome an evolutionary "experiment." Given such an enormous pool of horizontal within-species genome variation, plus the minuteness of steps involved in evolutionary change, the evolution of complex forms becomes less of an enigma and more of a tractable scientific problem that will no doubt someday be solved by machine computation. To invoke supernatural entities, "intelligent designers," or whatever because we have not yet elucidated the details of a complex evolutionary process or have not yet understood the details of a complex system is blatant junk science.

And any teaching of such junk science as "science" in the public schools is a disservice to the public and a dangerous incursion of religion into public school education. Because of its claimed and media-publicized connection to real science, the danger of intelligent design theory to science education has been taken more seriously by scientists than "young Earth" creationism. In

2002, the American Association for the Advancement of Science (AAAS) published a resolution against intelligent design theory, the resolution pointing out that individual scientists and philosophers of science have provided substantive critiques of intelligent design theory, demonstrating significant conceptual flaws in its formulation, a lack of credible scientific evidence, and misrepresentations of scientific facts. The resolution emphasizes that the proponents of intelligent design theory have failed to offer credible scientific evidence to support their claim that the theory undermines the current scientifically accepted theory of evolution, and that the intelligent design movement has not proposed a scientific means of testing its claims. The resolution finally calls upon its members to assist science educators in understanding the "inappropriateness of intelligent design theory as subject matter for science education."

INTELLIGENT DESIGN CREATIONISM IN THE SCHOOLS

AFTER THE U.S. SUPREME COURT ruled in 1987 that teaching creationism is illegal teaching of religion in the classroom, creationists made a special effort to provide a scientific basis for their beliefs, and the intelligent design movement was one of the results. Many proponents of intelligent design theory protest that the theory is not creationism. But although the theory is certainly not young-Earth creationism, what is an "intelligent designer" but a Creator, and what is the proposition of a Creator but a form of creationism?

An illustration of the corruption of public school teaching of biology is provided by the experience of the schools of Dover, Pennsylvania, a small town in the southeastern part of the state. In December of 2004, the local school board ordered the eight high school science teachers to read to their biology classes a statement that attacks the theory of evolution and that promotes intelligent design theory as an alternative explanation for the origin of life. The statement also prohibits scientific discussion in the classroom of the origin of life, saying the topic should be discussed by "individual students and their families."

This was apparently the first time intelligent design theory was incorporated into a U.S. public school curriculum. The science faculty protested, refused to read the statement, and instead, in January 2005, the Dover school superintendent and his assistant visited every ninth-grade biology class to read the four-paragraph statement himself without taking any questions afterward.

The statement says in part:

Because Darwin's theory is a theory, it continues to be tested as new evidence is discovered. The theory is not a fact. Gaps in the theory exist for which there is no evidence. A theory is defined as a well-tested explanation that unifies a broad range of observations. Intelligent design is an explanation of the origin of life that differs from Darwin's view. The reference book, Of Pandas and People, *is available for students who might be interested in gaining an understanding of what intelligent design actually involves.*

Elsewhere, in a press release, the Dover school superintendent says: "The Dover Area School District supports, and does not discriminate against, students and parents who have competing beliefs, especially in the area of the origin of life debate. The school board has noted that there are opinions other than Darwin's on the origin of life."

The above is a remarkable exposure of confusion. Darwinism concerns the origin of *species*, not the origin of life—the origin of life is a specialized field in modern biology and astrobiology more or less independent of evolutionism. If the idea is that evolutionism needs correction, there is no reason to prohibit discussion of the origin of life, unless the goal is to prevent any discussion of the origin of life except as an origin due to a Creator. The motives of the school board are clear: They do not want the children of Dover to be exposed to any science that contradicts their religious views. Since they cannot, by law, prohibit the teaching of evolutionism, they introduce intelligent design theory as a supposed "correction," and then prohibit discussion of the origin of life, a prohibition not yet struck down by the courts.

The book *Of Pandas and People* mentioned in the statement read to students refers to sixty copies of a book explaining intelligent design theory, the copies donated by an unidentified person to the school district.

Proponents of intelligent design theory who say their effort is merely an attempt to amplify one science (evolutionary biology) with more science (intelligent design theory) are doing the public a great disservice by distorting the controversy. All of the effort is an attempt by religionists to twist the teaching of science to children in the public schools in order to preserve certain religious views of parents.

CREATIONISM IS A PRODUCT OF RELIGIOUS BELIEF

LET'S CONSIDER A PARALLEL EXAMPLE. The term "numerology" refers to the study of the occult meanings of numbers and of their supposed influence on human life. Imagine a group of Americans who claim that numerology should be introduced into the study of arithmetic in public schools on the basis that mathematicians have not considered the full extent of what they study. Pressure is brought to bear by numerologists on local school boards, and since most members of local school boards are not mathematicians or knowledgeable about number theory, what will happen? The likely outcome is that local school boards will seek the advice of mathematicians, asking if there is any validity in numerology and should it be taught in arithmetic classes in public schools. Most mathematicians will snicker and say no, the school boards will tell the numerologists to go home, and that will be the end of it.

So what is different about creationism? The difference is that creationism is a product of religious belief, which in turn is a product of humanity unsettled by knowledge of mortality and the apparent indifference of the cosmos to human suffering. Religion gives comfort to people, which is why it has survived in one form or another for so long. When creationists approach or infiltrate school boards, they are essentially telling these school boards, "Look, science is in conflict with our Christian religion, and it makes us uncomfortable, and it should make you uncomfortable too, since we know that you also are Christian, and we need to be careful about indoctrinating children with materialistic evolutionism, which denies God and is dangerous to our family values and our country. What we need to do is introduce counterarguments into the teaching of evolution so that our children are not driven from religion, and if not counterarguments then at least tell children that evolutionism is not the only answer to the riddle of life. We would like to see the teaching of evolutionism banned, but we'll settle for counterarguments."

When such ideas are presented by creationists to school boards, do the school boards immediately admit their own lack of expertise about evolution and turn to scientists for an assessment of creationism? Apparently not often enough. According to surveys, a majority of Americans do not accept evolution as the explanation for life on Earth, and one can assume that a majority of people on school boards are of the same opinion, particularly in the "Bible Belt" of the South. That is the reason that creationists have been able to be politically successful where numerologists would certainly fail: Creationists have a view of divine power and of the origin of man in common with the members of school boards. The outcome, unless prevented by sophisticated political pressure and media attention, will ultimately be another "tyranny of the majority," in this case, a majority with an antiscience bent because such a bent provides personal comfort. Reason is not the issue here; the issue is personal comfort.

A further problem to that found in schools concerns creationist pressure to twist the science presented by science museums. Such museums are an important element in the education of the public, particularly in the education of children. These museums show a great many films about science, and films that mention the subject of evolution or the big bang theory of cosmology or films that provide information about the geology of the Earth are being turned down by many science museums as potential causes of trouble with local creationist groups.

The result, of course, is that many film producers hesitate to make such films because of a reduced market for them among science museums. What happened at the Fort Worth, Texas, Museum of Science and History is an illustration of what is happening throughout the country. A film called "Volcanoes," released in 2003 and sponsored in part by the National Science Foundation and Rutgers University, was turned down after the museum screened the film to a sample audience of 137 people. The film is about deep sea vents, the exotic living forms that have been discovered there, including primitive thermophilic bacteria that may reveal important information about the early evolution of life on Earth. Some members of the screening audience said the film was "blasphemous."

The marketing director of the museum apparently turned the film down for fear of creating controversy because it offended some people. A similar refusal to show the film occurred in approximately a dozen other science centers, many of them in the South. A film about Darwin's voyage to the Galapagos Islands was similarly turned down by a number of museums because it deals with evolution. A film called "Cosmic Voyage," which presents the universe in a scaling perspective from subatomic particles to clusters of galaxies, was refused by museums because it mentions the big bang theory of cosmology.

What is happening is that creationists are directly or indirectly controlling the education of the public to suit their religious views. Apparently, to the marketing director of a museum that may receive both state and federal funding, if science conflicts with public religious myths, science must be sacrificed. Decisions about science films in museums are being based not on the merits of the films as presentations of science, but instead are being based on whether or not the films will offend creationists.

SHOULD WE TEACH CHILDREN REALITY OR MYTH?

IS THE WHOLESALE TWISTING OF science to suit religion tolerable in public education? Beyond creationist attempts to force the teaching of intelligent design theory, and the pasting of antievolutionism stickers on textbooks, a cloud of fear has pervaded American schools. In many public school districts throughout the United States, the intimidation of science teachers is currently rampant: Many teachers simply ignore teaching evolution in order to avoid "trouble with the principal if word got about" that a teacher was teaching evolution to her students. Has any real progress been made since the infamous Scopes trial of 1925, the trial that had a Tennessee schoolteacher accused and dishonored for teaching evolution?

Our question concerning the ten-year-old boy and his sister in the heartland is not yet answered by the American public. What shall we teach these children in school? Shall we teach them the myths and lies of creationism because they give us comfort? Or shall we arm them with the knowledge of the world that science provides?

15

STEM CELLS: THE PETRI DISH BLUES

In this age in which everything is held to be permissible as long as it is freely done, and in which our bodies are regarded as mere instruments of our autonomous rational will, repugnance may be the only voice left that speaks up to defend the central core of our humanity.

— Leon Kass

Anesthesia was once decried as immoral because it contravened God's will for women to suffer during labor, and in vitro fertilization was initially assailed as a dehumanizing slippery slope.

— Christian P. Erickson

THESE DAYS LUDDISM TAKES strange forms. According to the chairman of the President's Council on Bioethics, the true test of whether scientific research should be permissible depends on whether or not the research seems repugnant. The former chairman, Leon R. Kass, has called this "the wisdom of repugnance." (Kass stepped down as chairman of the Council on October 1, 2005.)

But whose repugnance? And where and in what time? Evidently it's the members of the President's Council on Bioethics who decide what is repugnant and what is not repugnant. That council and something called the Bioethics Project, a group organized by conservative magazine editor William Kristol, son of conservative authors Irving Kristol and Gertrude Himmelfarb, rule bioethics in America today. The neocon Luddite cabal of philosopher Leon Kass, newspaper columnist Charles Krauthammer, magazine editor William Kristol, and social theorist Francis Fukuyama has decided that some areas of scientific research (including research on extending human life) are too disgusting to be tolerated and require prohibition. Since Leon Kass was appointed by President George W. Bush to chair the President's Council on Bioethics, one can assume that President Bush is in agreement with the attitudes of this cabal. The regulations enacted by the Bush administration forbidding federally funded embryonic stem cell research are certainly in accord with those of the neocon Luddites.

Although this bio-Luddite cabal contains no scientists, both Kass and Krauthammer have had some exposure to science, since they were at one time medical students. But whatever science they absorbed has apparently never been a factor in their arguments, which are for the most part based on their views of ethics, morality, religion, and an elitist attitude that they are the appointed shepherds of the flock. Where are we then if not back to the medieval theological councils issuing edicts about what can and cannot be done by people in the sciences? The early medieval mind found the study of human anatomy repugnant, and for a thousand years the progress of human medicine stagnated. Shall we accept the "wisdom of repugnance" as a guide for scientific research?

STEM CELLS, EMBRYOS, AND TALKING HEADS

THE EMBRYONIC STEM CELL DEBATE is like many other debates in American life: Nearly everyone has an opinion, even people who know nothing about the subject. In the midst of the debate, a television talking-head conservative commentator opines with gravitas that using embryonic stem cells for research is immoral because such cells contain a complete copy of the human genome. Since every cell in the human body, except certain cells of the immune system, contains a complete copy of the human genome, maybe the television commentator does not understand that the relevant difference between embryonic stem cells and other cells is not which cells contain complete copies of the human genome—they nearly all do—but merely which cells have certain genes turned on or off. Conceptually, this is grade-school stuff, only a modicum of homework required before going on the air, just a little homework to prevent perverting the public debate with misinformation and junk science.

Stem cells are cells whose gene activation profiles—which genes are turned on or off—are not yet fixed to make them specialized cells. In some stem cells, the gene profile is partially fixed, and they are called "pluripotent" stem cells. In other stem cells, the gene profile is not fixed at all, and they are called "totipotent" stem cells. Early-stage embryonic stem cells are totipotent stem cells, the gene profile totally unfixed, so they have the potential to become any type of cell at all.

The term "fixed" needs to be qualified: The fixation can be reversed if the genome is exposed to certain circumstances. In cell biology, the term "differentiation" is used to denote complete or partial fixing of the genome. If you like computer program analogies, the genome is a collection of subroutines, some activated, others inactivated. The subroutines activated in a muscle cell are different than those activated in a liver cell—but both cells contain the whole collection of subroutines, the entire genome of that particular individual. Under ordinary circumstances, once a muscle cell, always a muscle cell—a muscle cell ordinarily cannot be turned into a liver cell and vice versa. Stem cells, however, are cells that are not yet fully differentiated, and embryonic stem cells from the blastocyst stage are able to differentiate into *any* type of cell. Theoretically, when we learn the necessary conditions, we will be able to make embryonic stem cells differentiate into nothing but muscle cells, or liver cells, or any kind of cell we choose.

Maybe the debate about embryonic stem cells would have traveled in a more sensible direction if biologists had not been so unclever with their public terminology. Although most people use the words "embryo" and "fetus" interchangeably to refer to the entire gestation period between conception and birth, in humans the term "embryo" refers technically only to the stage of development between two weeks and eight weeks. During the first two weeks after fertilization, the entity is technically a blastocyst. Between two weeks and eight weeks, the entity is an embryo. After eight weeks and until birth, the entity is called a fetus. Totipotent stem cells, so-called embryonic stem cells, are actually blastocyst stem cells in an undifferentiated mass of cells less than two weeks old, at a stage before any differentiation into an embryo has begun. Had such cells been called "blastocyst stem cells" instead of "embryonic stem cells," the public might have been better served. As it happened, the idea of tearing human "embryos" apart for research drove some people to the ramparts.

MAYBE THE MAJESTY OF LIFE NEEDS
SOME IMPROVEMENT

THE CENTRAL FACT IN THE debate concerning embryonic stem cells is that blastocyst stem cells in all early embryos and adult stem cells in some tissues are undifferentiated cells that in response to appropriate signals differentiate and give rise to a variety of cell types.

That is the basic story of embryology, the story of the biologically automated engineering of any animal or any human being. Before the first half of the twentieth century not enough was known to even guess at what was happening at the level of genes, cells, and tissues. By the end of the twentieth century, there were still many puzzles, but the great riddle of embryology was essentially solved and the solution commonplace to biologists throughout the world. Of course, one predictable consequence of solving this great riddle was that philosophers and journalists and religionists rose up in dismay with complaints that science was destroying the mystery and majesty of life. In all the brouhaha, no one paid much attention to the pleas of scientists that only by demystifying life could we save life when people needed their lives to be saved. As for the majesty of life, the truth is when biomedical scientists think about smallpox, leprosy, cancer, heart disease, Alzheimer's disease, Parkinson's disease, plague, Spanish influenza, childhood leukemia, and all the other biological scourges, the "majesty" of life appears to need some improvement.

Understanding how cells differentiate into tissues and organs will improve the majesty of life rather than diminish it, and it will ultimately, in what we can hope is the several billion year course of human existence on this planet, save more lives than you or I can possibly imagine.

DIFFERENTIATION IN DEVELOPMENT

DIFFERENTIATION IS THE KEY TO understanding development. Again, except for the immune system, no matter what the cell type, all of the genes defining the human organism are present in every cell nucleus: The variations between tissues are due not to the presence or absence of certain genes, but to the expression of some genes and the repression of other genes. For example, one of the best understood cases of cell differentiation is that of the red blood cell, the major protein of which is hemoglobin, the protein that binds and carries oxygen.

The hemoglobin gene is also present in the cells of all other tissues, but in these tissues the gene is barely or never expressed. In contrast, in the DNA of the developing red blood cell, "enhancer" nucleotide sequences have been identified in and around the regions encoding hemoglobin, and these enhancers apparently interact with regulatory proteins specific to red blood cells, the result the full-scale transcription of hemoglobin genes only in those cells. (In this context, "transcription" is the process by which genetic information in DNA is converted into RNA, with the RNA ultimately "translated" into protein.)

We are what we are, each of us, because of the chemical programming in place when the fertilized egg cells of our mother, now incorporating both maternal and paternal DNA, began the choreographed ballet of cell division and differentiation that ultimately led to a new human being. Most of that chemical programming is immortal, passed from one generation to the next, and not only immortal but universal, the common treasure of every human being on the planet. Does an understanding of this grand scenario diminish the majesty of life? Of course not.

In humans, there are approximately two hundred different types of cells organized into tissues, with each tissue typically consisting of an organization of a small number of cell types. The ultimate production of various cell types from the original fertilized egg cell constitutes differentiation, and the biochemical basis of cell differentiation is the synthesis by cells of particular sets of proteins, carbohydrates, and lipids. These syntheses are catalyzed by special proteins (enzymes), and each enzyme is in turn synthesized in accordance with the instructions coded in a particular gene, a sequence of nucleotides in the DNA of the cell nucleus. Such is the grand scheme that links biochemistry, molecular biology, genome biology, cell biology, and embryology.

We are complicated animals, and the various tissues that form our organs are an important part of our complexity. Questions about tissues and organs were asked by scientists for centuries, but no answers came back from the laboratories and the riddles remained unsolved. Unfortunately, now that we have finally reached the edge of a new domain of knowledge, commissars of ethics present us with diktats about "repugnance" and forbid us to go there.

EMBRYONIC STEM CELLS AND BLASTOCYSTS

THE ULTIMATE TOTIPOTENT STEM CELL is the original fertilized egg cell, completely undifferentiated, whose progeny progressively differentiate and replicate into an entire adult human. Embryos have many types of stem cells, totipotent and pluripotent, and during embryological development these stem cells are triggered to differentiate into various types of tissue cells. In adult organisms, only certain tissues (for example, skin and bone) contain stem cells, and these cells are involved primarily in cell-replacement functions.

Let's consider some scientific details. Human embryonic stem cells come from so-called preimplantation embryos, most of which are generated at in vitro fertilization clinics. Within days after fertilization, the "embryo" consists of a hollow sphere of cells, the blastocyst, which contains within it a cluster of a few hundred identical cells called the "inner cell mass" that can eventually develop into an embryo and then into a fetus. When removed from the blastocyst, these inner mass cells can be propagated indefinitely in specialized media—in a petri dish. When the medium around the cells is changed to allow differentiation, the cells continue to divide and aggregate in the petri dish to form "embroid" bodies. Although these cell clusters lack the organization of an embryo, the clusters have differentiated in the petri dish into various tissue types, including skin, muscle, bone, and neurons.

We are cracking riddles here, beginning an adventure that will alter the course of human existence. No one alive today will see the adventure play out, but everyone alive today is part of it, since this is an adventure of the human species.

Given that many diseases involve defective cell types or injured tissues, possible therapeutic uses of stem cells are a natural and important focus of attention. Clinicians have exploited stem cells for therapeutic purposes for more than forty years. Bone marrow transplantation, for example, is lifesaving stem cell transplantation for patients with certain types of bone marrow diseases and malignancies. But the usefulness of stem cell transplantation of other cell types was in the past limited by the fact that many organs (for example, brain, spinal cord, heart, kidney) were believed to lack detectable stem cells. It was also believed that cells from these organs could not be reprogrammed to differentiate into different cell types during adulthood. But that has all changed.

Three recent discoveries have revolutionized stem cell biology and have demonstrated the clinical potential of stem cells in a wide range of human diseases: First, stem cells have been detected in organs previously thought to lack stem cells and regenerative potential. Second, organ-specific adult stem cells appear to display much more plasticity than previously thought, and stem cells isolated from one tissue can differentiate into a variety of unrelated cell types and tissues. Third, human embryonic stem cells can now be isolated from blastocysts and made to differentiate in vitro into a wide variety of cell types. Totipotent embryonic stem cells are derived from the inner cell mass of the blastocyst, a stage so early that no distinctive human or animal form can be recognized in the cell mass. What is there is the potential for the formation of a human being, but in principle, given the proper conditions, every cell in the human body has the "potential" to generate a complete human being—in the way that an acorn has the potential to generate an oak tree.

The discovery of stem cells in adult tissues, the unexpected plasticity of both adult stem cells and differentiated cells, and the isolation of human embryonic stem cells have expanded the prospects for stem cell–based therapies and created an entirely new field in biomedicine. The consensus in the medical community is that stem cell–based therapies using autologous donor cells (cells from the same individual) have tremendous promise for the treatment of both acquired and inherited diseases involving tissue degeneration and cellular dysfunction.

While researchers have been working in the background on stem cells for years, it took two publicized breakthroughs—animal cloning by Ian Wilmut and his group in 1997, and the derivation of pluripotent human embryonic stem cells by James Thompson and his group in 1998—to bring the entire field to public attention. The Wilmut group demonstrated that an adult cell nucleus can be reprogrammed to produce an entire animal, and the Thompson group provided a possible source of cells for cell-based therapies for many human diseases. Even more intriguing was the notion of combining the two approaches in the process of "therapeutic cloning," whereby embryonic stem cells could be derived (at least in theory) from the cells of a patient, thus avoiding problems of immune rejection, and then coaxed into forming large quantities of the cell types needed for a cure.

I'VE PROVIDED THESE DETAILS of the biology of human stem cells in order to make a point: For the most part, researchers in this field are not in a state of sudden philosophical shock about their discoveries. Our current knowledge of stem cells in embryogenesis and early development has required two generations of hard work, pieces of information accumulated one after the other in hundreds of laboratories, and insights piled on insights over the years. It was only when the work became publicized during the 1990s that philosophers, ethicists, moralists, religionists, and ultimately politicians began barking their objections based on ideas of human dignity and "repugnance," objections to a modern scientific enterprise (human embryology) ongoing for more than a century. Although the basic objections are not against the study of embryos but against the manipulation of embryos, the successful study of any system requires manipulation of that system.

So what is special about this case? What is special is apparently the religionist idea that the formless mass of a few hundred cells constituting the inner cell mass of a blastocyst is already a human being. I call this a "religionist" idea because there is absolutely no basis in science for this belief, and any argument for the idea that portends to be scientific is junk science. Although they are both "embryos," an acorn is not a tree and a blastocyst is not a human being.

THE OBJECTIONS TO EMBRYONIC STEM CELL RESEARCH

NO PUBLIC COMMENTATOR HAS BEEN more vigorous in opposition to stem cell research than Charles Krauthammer, a conservative syndicated newspaper columnist and member of the Bush President's Council on Bioethics. Krauthammer's basic premise is that the human embryo should be an inviolable entity, and with that as a premise he proposes that research on embryonic stem cells will result in "the deliberate creation of embryo factories for the sole purpose of exploiting and then destroying them." Associated with this, he sharply criticizes arguments about potential near-future cures with stem cells as "a cruel deception perpetrated by cynical scientists and ignorant politicians."

Technically, Krauthammer concerns himself with the stage from the fertilized ovum through the blastocyst stage (ovum-morula-blastocyst) rather than with the embryo stage, but we'll use the terms "embryo" to denote all the entities. Let's consider the premise, the proposition, and the associated criticism of the idea of potential stem cell therapies.

The premise is clear: Krauthammer (and others) believe that beginning with conception (fertilization of the ovum), the embryo must be inviolable. But like any premise in an argument, this premise can be examined critically, and the critical question is: Why? What are the reasons for why the embryo must (should) be inviolable? And if the embryo must be inviolable, why not the fetus, the infant, the child, and the adult? And of course these questions bring us out of the arena of science and into the arenas of politics and ethics. As far as science is concerned, there is no basis for assigning inviolability to any stage of human existence. Does that mean that scientists believe all stages of human existence can be violated? Of course not; they usually believe just the opposite. What it means is that the justification for the assignment of inviolability cannot be science but must be political or moral—the justification cannot be defended with science or proposed as the outcome of scientific analysis.

Then what can science offer to the question of the violability or inviolability of the embryo? The answer is this: In principle, if the chemical environment of the fertilized ovum-to-blastocyst stage (we'll call this the "early embryo" stage) is manipulated, the early embryo will not develop into a human being at all. With a blastocyst in a petri dish, given our control of the chemical environment in that petri dish, it is we who in principle control the development of that blastocyst. In principle, we could make the blastocyst stem cells produce only skin cells; not a human being, just a mass of human skin cells.

Is the clump of blastocyst stem cells a human being? Do they have the potential of becoming a human being? So do human sperm cells and human egg cells, most of which are never allowed to do so. If a woman produces 400 egg cells during her lifetime and has only two children, 398 of her egg cells have never achieved their potential for developing into an individual human. Of course, you may want to consider those unfulfilled egg cells as sanctified, but that's a religious position and not a conclusion deduced from science. Similarly, the few hundred blastocyst stem cells in each early embryo are merely the result of unfulfilled egg cells that have been fertilized and allowed to undergo repeated cell division a half dozen times or so.

And that's it. What science suggests is that the blastocyst stage is not yet a stage of human development, it's a cell mass not yet differentiated, and it's the degree of differentiation which should be the important parameter in determining whether any entity in a petri dish is or is not a "human being." The blastocyst is not a human (or even animal) of any kind: no nervous system, no brain, no mental capacity, no organs, nothing but a clump of cells without organization. In short, I believe Krauthammer's premise of the need for inviolability of the early embryo (blastocyst stage) to be without any rational basis. An undifferentiated blastocyst is a collection of cells carrying the human genome, but it's not a human being at all.

With the premise out of the way, let's consider Krauthammer's proposal that research on embryonic stem cells will result in "the deliberate creation of embryo factories for the sole purpose of exploiting and then destroying them."

Absent the premise that the embryo (blastocyst) must be inviolable, the objection to "exploitation and destruction" of the embryo becomes vacuous, no more reasonable than an objection to research on Henrietta Lacks's immortalized tumor cells, the cells that carry Henrietta Lacks's complete genome (see below). Tumor cells, in fact, have undergone partial reversal of differentiation. And that's the point: The degree of differentiation is the key to the difference between human actuality and mere potentiality. The clump of the inner mass cells of the blastocyst is no more a "human being" than a clump of human tumor cells is a human being.

When a woman named Henrietta Lacks died in 1951 of a cervical carcinoma, some of her tumor cells were removed and continuously cultured down to the present as an immortal strain of carcinoma tumor cells, the so-called HeLa cells widely used in cancer research. Since Henrietta Lacks's immortal tumor cells contain the deceased Henrietta Lacks's entire genome, we could in principle (however unlikely) recover (reestablish) an appropriate gene profile in one of her tumor cells that would begin the development of a copy of Henrietta Lacks as an embryo, fetus, infant, child, and ultimately an adult—in other words, an unlikely but theoretically possible "clone" of Henrietta Lacks. Will the Kass Council on Bioethics consider banning research on human tumor cells because such cells can in principle serve as the origin for the genesis of a human being? That reasoning is apparently the basis for the ban on research on human embryonic stem cells. And the same reasoning could be applied to research involving the cells of any human tissue.

Finally, Krauthammer's criticism of public statements by biomedical scientists and others concerning potential near-future cures with stem cells are just off-base. Krauthammer has called these arguments "a cruel deception perpetrated by cynical scientists and ignorant politicians."

Let's forget the rhetoric (this book is replete with my own rhetoric) and focus on Krauthammer's idea that there are no near-future prospects for stem cell therapy in human medicine. It's an odd idea for someone with a medical background (even if in psychiatry), since the current biomedical literature contains hundreds of published studies of stem cell therapy in animals, and one can as-sume that the thousands of researchers involved in such studies are not devot-ing time and energy to these studies because they anticipate failure in applications of the animal studies. The scientific and medical communities, as evidenced by the scientific and medical literature, obviously think the research worth pursuing, so why not Krauthammer? Certainly he must understand that the absence of evidence from human trials is not important at this stage of the science, since human trials are never the first to occur but always the last stage of medical research.

Krauthammer says:

> Stem cell research does hold promise for clinical cures in the far future. But right now we're at the stage of basic science: We don't understand how these cells work, and we don't know how to control them. Because their power is so extraordinary, they are very dangerous. Elementary considerations of safety make the prospect of real clinical application distant.

Now Krauthammer's entire argument crumbles into sophistry. First, with an unreasonable assumption of moral inviolability, Krauthammer argues that em-bryonic stem cells should not be created, exploited, and then destroyed. And even if they are created and exploited, Krauthammer argues that the exploita-tion is useless and dangerous because the applications of stem cell research are a matter for the "far future." Krauthammer, it seems, has been peering into a crystal ball and he's advising us where human stem cell therapy will be in five years, ten years, twenty years, or maybe even one hundred years. Krauthammer says the research will go nowhere. He says stem cell research will go nowhere until the far future. The "far future" is the far future. The problem is that the future of science is unpredictable—unless scientific research is shut down by government fiat.

Which brings us to the culmination of Krauthammer's antiscience mani-festo, as he proposes, to the conservative readers of The Weekly Standard, a strat-egy for opposing stem cell research, "One must not underestimate the efficacy of political restraint. If you can restrain for decades something that promises a cure, imagine how many other, less morally repulsive, substitute cures will present themselves in the meantime. You cannot stop evil science, but you can delay it, and thus possibly supplant it."

Krauthammer's strategy, and the strategy of the Bush President's Council on

Bioethics, of which Krauthammer is a leading member, is thus to cripple scientific research because of a personal hysterical view that such research is evil. We have apparently returned to a medieval mode of dealing with the world.

JUNK SCIENCE IN THE BIOETHICS COMMISSION

WHERE ARE WE IN BIOMEDICAL research on stem cells? The prohibitions recommended by the Kass council and signed into law by President Bush have effectively crippled the stem cell research program in the United States. Many American scientists believe that the President's Council on Bioethics has made public bioethics the servant of politics. The president's stem cell policy is based on junk science and has severely curtailed opportunities for American scientists to study cell lines that have unique attributes or represent invaluable models of human disease. This is government working against the public good for political advantage with religious groups.

In truth, the junk-science–based refusal of some segments of American society to accept embryonic stem cell research is a remarkably vicious attack against their own children and grandchildren. Ultimately, after the current medieval mood passes, stem cell research and consequent stem cell therapies will be a vigorous part of American biomedicine, and Americans will be catching up with countries that have gone forward with the research. How many Americans, including children, will have been killed by the delay?

Four years have passed since August 9, 2001, when President George W. Bush announced that research on human embryonic stem cells created before that date would be supported by federal dollars, and research on lines created later would not. There is apparently no disagreement that the president's policy has severely curtailed opportunities for American scientists to study cell lines that have since been established in other countries, cell lines many of which have unique attributes or represent invaluable models of human disease. The Kass-Kristol-Krauthammer-Fukuyama cabal has been successful.

If the American ban continues, cell therapies derived from embryonic stem cell research will be developed in those countries that are now permitting such research, and eventually rich Americans with relevant treatable diseases will travel to those countries to receive stem cell therapy. The future score sheet resulting from the American ban is plain: Rich Americans will live; other Americans will die.

Of course not everyone cares about the dying. When neoconservative Francis Fukuyama was asked if the government has the right to tell an ordinary citizen (not a criminal) that he must die, Fukuyama responded: "Absolutely." Another neoconservative, Eric Cohen, has proposed that health and excellence are "gifts"; some people have them and some don't. Are these totalitarian and fascist ideas the ideas we want to be the basis for our future?

Approximately 130 new human embryonic stem cell lines have been produced worldwide in other countries since the president's decision to restrict embryonic stem cell research in the United States. In Singapore, researchers have cultured new lines uncontaminated by nonhuman animal products, new cell lines that are therefore preferable for applications in human patients to the contaminated old cell lines still available in the United States. Sweden has approved embryonic stem cell research, and so have China, Japan, India, Israel, Australia, South Korea, Switzerland, Brazil, Italy, and several other countries—all forging ahead with research in this new field of biomedicine.

THE BUSH ADMINISTRATION'S EMBRYONIC stem cell policy is a sad mistake. History will note that four members of the Kass Council on Bioethics—all scientists— opposed the council's recommendation to President Bush to ban federally funded embryonic stem cell research. The strategy of neoconservative bio-Luddites is to impede or block completely any scientific research that they find "repugnant," and at the moment the rest of us are all victims of this socially destructive, antiscience attitude that is certainly more dangerous to the public than any cluster of cells in a petri dish. Leon Kass's "wisdom of repugnance" as a guide for science policy is a sophomoric folly.

16

CLONING: WILL WE HAVE
A HUNDRED JERRY FALWELLS?

> In the three short decades between now and the twenty-first
> century, millions of ordinary, psychologically normal people
> will face an abrupt collision with the future.
> —Alvin Toffler, on human cloning, 1970

> I don't like the idea of copying people.
> —Ian Wilmut, biologist who copied (cloned) the sheep
> Dolly in 1997

MORE THAN FORTY YEARS ago, an editorial in *The New York Times* declared, "Geneticists are on the threshold of an historical breakthrough in their efforts to probe the secrets of heredity. Is mankind ready for such powers?"

These days, the "secrets" of heredity are taught to high school students, and mankind has absorbed modern genetics to the point of commercializing genetics wherever and whenever it can. The world goes on, usually absorbing each revolution in science, applying the new knowledge and moving forward. But every scientific revolution is a shock of sorts to society, and in recent years no shock has been more potent than the idea of making "exact" copies of animals and human beings—cloning.

The popular understanding of cloning has its roots in Alvin Toffler's 1970 book *Future Shock*, in which Toffler took a clear scientific concept and twisted it into the junk-science fantasy prediction that man will be able to make biological carbon copies of himself. Unfortunately, this fictitious version of cloning was presented in a highly influential nonfiction book, and clones suddenly morphed from the simple progeny of asexual reproduction into sophisticated products of biological engineering created by scientists bent on controlling nature.

During the past thirty years, the media has consistently produced twisted science in its reporting of cloning, and the result is a frenzied public junk-science circus around the subject. The danger to the public is great: Cloning of animals and humans will be a subject of ever-growing importance during this century, and the public needs to understand its positive and negative consequences.

CLONING HAS MANY FORMS

IN SCIENTIFIC PARLANCE, "CLONING" IS a broadly used shorthand term that refers to producing a copy of some biological entity—a gene, an organism, a cell—a result that in many cases can be achieved by means other than the technique known as "somatic cell nuclear transfer," the technique involved in the creation of the famous sheep, Dolly. Bacteria clone themselves by repeated fission. Plants reproduce clonally via asexual means and by vegetative regeneration. Somatic cell nuclear transfer, used to produce the sheep Dolly and discussed in detail later, is simply the replacement of the nucleus of an egg cell with the nucleus of an ordinary cell, a non-germ cell (somatic cell). The new nucleus is reprogrammed by the egg cell, and when the egg cell divides repeatedly to ultimately produce a new individual, that individual has the genome of the new nucleus and not the genome of the original nucleus. So if the new nucleus came from a somatic cell of the dog Fido, and it's implanted into the egg cell of the dog Rover, successful embryonic development produces an exact copy of the dog Fido. The nucleus carries the genome; change the nucleus and you're changing the genome; the genome determines the individual. (This is exactly what occurred recently when the first dog was cloned: A nucleus of one breed was implanted into the egg cell of another breed to produce a clone of the first breed.)

Broadly defined, cloning of a complete organism is asexual reproduction that results in progeny genetically identical to the parent. To many people, cloning was invented with the birth of the sheep Dolly. In fact, cloning has been practiced for millennia with plants, and for decades with mammals, and Dolly's birth followed an orderly progression of experiments that started with cloning frog embryos in the 1950s and mammalian embryos in the 1970s.

Crude cloning techniques have been in use for centuries. The practice of taking cuttings is universal among gardeners, and agricultural companies now propagate desirable plant strains in large quantities. Lower invertebrates can be easily cloned: For example, if one cuts an earthworm or flatworm in half, the halves will regenerate to create two genetically identical individuals. Although this method does not work in vertebrates, identical twins are naturally occurring genetic clones, and the method of nuclear transplantation, first used forty years ago in frogs, has been successfully used to make clones of various mammals, and could probably be applied to humans.

The first successful mammalian cloning by nuclear transfer, in which cells from cleavage-stage sheep embryos were fused with unfertilized sheep eggs, was reported in 1986. Successful cloning from older embryos (and ultimately from an adult cell, in the case of Dolly) challenged conclusions from previous work that the nuclei of differentiated cells are unable to support normal development.

It's important to keep in mind the distinction between the two major mammalian laboratory cloning techniques. Embryonic clinical research cloning (of importance in the stem cell debate) involves using early embryonic stem cells, for example blastocyst cells, to generate other stem cells for research and therapeutic purposes. One source of early embryonic stem cells consists of early embryos discarded in in vitro fertilizations.

In contrast to embryonic clinical research cloning, nuclear transfer reproductive cloning (of importance in the human cloning debate) involves implanting the nucleus of an ordinary tissue cell into an egg cell that has had its own nucleus removed and implanting that egg cell in a uterus to eventually produce an entire individual.

A third variant is nuclear transfer clinical research cloning. In this case, to avoid compatibility problems, instead of using discarded in vitro fertilization embryos as a source of blastocysts and stem cells, a patient's nucleus is fused with an enucleated egg cell, which begins development, and the consequent blastocyst is used as a source of stem cells for therapy with the donor of the nucleus.

NUCLEAR TRANSFER REPRODUCTIVE CLONING

A LARGE BODY OF EVIDENCE supports the notion that the spectrum of genes expressed in eukaryotic cells (cells that contain a nucleus, which means at least all animal cells) is influenced by the cytoplasmic environment in which the cell nucleus resides. The technique of nuclear transplantation has been especially useful in demonstrating this point. In one type of transplantation study, inactive nuclei have been taken out of cells that are synthesizing neither DNA nor RNA and transplanted into egg cells whose nuclei were previously removed. When placed in this new cytoplasmic environment, a nucleus that had previously been inactive will begin to synthesize RNA and, in some cases, DNA. In the most dramatic experiments of this type, carried out on the African clawed toad *Xenopus laevis*, John Gurdon removed the nuclei from cells lining the intestine of swimming tadpoles and transplanted them into egg cells whose nuclei had already been removed. Although the transplanted nuclei were derived from intestinal cells, they were able to direct the development of the egg cell into a complete new toad. Clearly the egg cytoplasm must exert profound effects on nuclear activity if it can induce the nucleus of an intestinal cell to behave in this way.

In 1997, Ian Wilmut announced that a sheep had been cloned using a somatic cell nucleus from an adult female sheep. This was the first time that an adult vertebrate had been successfully cloned from another adult. To do this, Wilmut and his colleagues took cells from the mammary gland of an adult (six-year-old) pregnant ewe and put them into culture. The culture medium was formulated to keep the nuclei in these cells at the resting stage of the cell cycle. They then obtained oocytes (the maturing egg cells) from a different strain of sheep and removed their nuclei. The fusion of the donor cell and the enucleated oocyte was accomplished by bringing the two cells together and sending electrical pulses through them. The electric pulses destabilized the cell membranes, allowing the cells to fuse together. Moreover, the same pulses that fused the cells activated the egg to begin development. The resulting embryos were eventually transferred into the uteri of pregnant sheep.

Of the 434 sheep oocytes originally used in this experiment, only one survived: Dolly. DNA analysis confirmed that the nuclei of Dolly's cells were derived from the strain of sheep from which the donor nucleus was taken. The conclusion was that the nuclei of adult somatic cells can become totipotent (capable of organizing differentiation into any cell type). No genes necessary for development had been lost or mutated in a way that would make the genes nonfunctional.

This result has been confirmed in cows and mice. In the mammalian ovary, cumulus cells constitute the mass of cells surrounding the egg cell in the follicle. In mice, nuclei from the cumulus cells of the ovary were injected directly into enucleated oocytes. These re-nucleated oocytes were able to develop into viable mice at a frequency of 2.5 percent. Cumulus cell nuclei from cows can direct the complete development of oocytes into mature cows. All of this is the reproductive cloning of animals.

THE CLONING DEBATE

THE MEDIA HAS OFTEN CONFLATED embryonic clinical research cloning and nuclear transfer reproductive cloning, a conflation that has only added to public confusion and made many scientists hold their heads in despair.

There are indeed some people who think conflation of the two issues is justified, the justification usually some version of a "slippery slope" argument proposing that if clinical research cloning is allowed, reproductive cloning is soon to follow. This is a counterfeit argument, as counterfeit as an argument in 1900 that research in aviation should be banned because if man learns to fly, some crashes to Earth are soon to follow. The argument, which is the same sort of argument made about many new technologies, is not cogent and not important except as a political device to twist public perception.

The slippery slope argument is also proposed solely within the context of clinical research cloning by declaring that allowing research on blastocyst stem cells will pave the way for research on fetuses and even infants. This argument is as specious as the slippery slope argument conflating clinical research cloning and reproductive cloning, but the argument is made nonetheless and with effect.

In general, it's not difficult to find scientific realities or sociological analysis twisted in the service of moral positions. The public records of the Bush Council on Bioethics headed by Leon Kass is as good a source of twisted science and reason as one can find.

In one meeting of the council, for example, the conservative syndicated columnist Charles Krauthammer, a member of the council who opposes all forms of cloning, stated the following:

> All of us were once blastocysts, and the question is, does a blastocyst deserve respect and why? One reason would be—I would call one mysterious and the other prudential. The mysterious would [be] because there is something unique and special about this particular collection of cells, namely that it can become us, there is something—it ought to put us in some kind of awe and some kind of position of granting it moral status and, therefore, practical or moral status and, therefore, treatment different from an ordinary collection of cells.

And further on, Krauthammer says:

> Were we to try to manipulate [the blastocyst], to play with it, to strip mine it, to use it, to manufacture it, it will predispose us to treating this kind of thing in a way so that we may break down the barriers toward treating a fetus that way, ultimately a child or a human that way. In other words, it is more an argument about us and the weakness of human nature and the temptations, particularly the temptations of technology than it is a statement about the nature of the cells itself.

So there we have it, the two main "moral" arguments against clinical research cloning: First, the human blastocyst is special, a mysterious entity of special status whose mystery should not be violated, and second, research on human blastocysts, if permitted, will ultimately lead to manipulation of fetuses, infants, children, and humans.

Slippery slopes are slippery slopes, and here from the minutes of one of the council meetings is an enlightening slippery slope provided by social theorist Francis Fukuyama, also a member of the President's Council on Bioethics:

> If some rich entrepreneur wanted to clone himself and give the embryos out as party favors, I mean it is a pretty gross thought but, you know, you could imagine some guy in Silicon Valley, you know, doing that as a joke to stick it to the antiabortion people. I mean, presumably even the, you know, supporters of that, you know, would not want to see that happen.

Thus is national policy made in the twenty-first century.

LIMITATIONS OF REPRODUCTIVE CLONING OF HUMANS

THE BELIEF OF SCIENTISTS THAT reproductive cloning of humans is not currently feasible derives from experience with animal cloning. Most cloned mammals derived by nuclear transfer die during gestation, and those that survive to birth frequently have the "large offspring syndrome," a neonatal phenotype characterized by respiratory and metabolic abnormalities and an enlarged and dysfunctional placenta. These animals die later. However, although most cloned animals die, enough survive and are normal to make cloning of certain agricultural animals commercially feasible, an implausible scenario in the case of human cloning.

But why is it so difficult to achieve viable mammalian clones?

In order for a donor nucleus to support development into a clone, it must be "reprogrammed" to a state compatible with embryonic development. Inadequate reprogramming of the donor nucleus is most likely the principal reason for the developmental failure of clones. The transferred nucleus must properly activate genes that are important for early embryonic development and must also suppress genes associated with differentiation that have been transcribed in the original donor cell. The cloning failures during development may reflect inappropriate expression of many genes whose harmful effect is exerted at different developmental stages. These fundamental limitations of cloning are a focus of analysis of the underlying cellular mechanisms, and in time this information may be useful in the development of clinical treatments involving the reprogramming of specific cell types. But so far, successful adequate reprogramming in mammals is rare rather than common.

Several laboratories have used a variety of somatic cell types to create cloned sheep, cattle, mice, pigs, goats, rabbits, cats, and most recently a dog. But cloning has failed in the rat and rhesus monkey. In all known cases of success, only a small proportion of embryos reconstructed using adult or fetal somatic nuclei develop to become live young, typically 4 percent or less. The low overall success rate is again apparently the cumulative result of reprogramming inefficiencies at each stage of the process.

Despite these experimental difficulties, the fact that reproductive cloning by nuclear transfer from adult somatic cells can be achieved at all is a major discovery. Technical problems exist, but reproductive cloning of animals (nuclear transfer cloning) will eventually be commonplace. In a study of twenty-four surviving sexually mature cattle successfully cloned from somatic cells in 1998 and 2000, results of physical examination were normal for all animals, including objective (temperature, pulse, respiratory rate) and subjective (general appearance, eyes, lymph nodes, and cardiac and pulmonary auscultation) findings. Results of abdominal palpation of reproductive and gastrointestinal organs and kidneys were normal. Social interaction and behavior of the cloned animals were normal. The researchers reported the animals had normal conditioned responses such as reacting to farm equipment used for feeding. The animals had developed a social dominance hierarchy and the full spectrum of behavioral traits. The cloned animals exhibited puberty at the expected age and body weight. Conception rates after artificial insemination were excellent, and two of the cloned animals gave birth to calves that appeared normal in every respect. It seems apparent that if there are technical problems in some places and with some species, the problems will eventually be solved and reproductive cloning of animals will be routine.

HUMAN CLINICAL RESEARCH CLONING

THERE ARE NO SCIENTIFIC ARGUMENTS against human clinical research cloning, which essentially involves the production of human blastocysts as sources of totipotent stem cells to be used in basic research and clinical applications. All the arguments involve subjective judgments about the "mystery" of blastocyst cells and "slippery slopes" that supposedly will lead to destructive research on embryos, fetuses, infants, and so on. The operative idea, as detailed in the previous chapter on embryonic stem cells, seems to be "repugnance," with an implicit belief that contemporary repugnance is a justification for prohibiting a new technology that may in the future save the lives of many thousands of children and adults.

Although research cloning and reproductive cloning are separate issues, many of the people opposed to both realize that it's in their interest to have the two issues conflated in the public mind, so that strong public rejection of reproductive cloning will carry along with it rejection of research cloning. It's an explicit strategy, and at the present time the supporters of clinical research cloning are at a disadvantage in the United States because they are not organized and because the public has been misinformed about clinical research cloning and the nature of blastocyst cells.

HUMAN REPRODUCTIVE CLONING

AS FOR REPRODUCTIVE HUMAN CLONING, the history of technology tells us that human clones will eventually be an ordinary part of the human population, without any of the ludicrous ideas about it now promulgated, ideas that involve objections that are primarily religious or aesthetic in content. It's probable that several hundred years ago, when the heart was considered the repository of the "soul," the idea that a heart could be transplanted from one individual to another would have nauseated some people and caused an uproar among ecclesiastics.

Putting aside the idea of inevitability, if one examines the reasons for and against human reproductive cloning that have been put forth by one side of the debate or the other, what is striking is that for the most part the opponents of human reproductive cloning are all against destabilizing or changing the current structure of society, while those in favor of permitting human reproductive cloning for the most part consider it an important step forward in human and societal evolution. It's all opinion, of course, since neither side can marshal much evidence to support its view. Fukuyama's fear of embryos as party favors in Silicon Valley is an example of the level of discourse.

The minutes and reports of the Bush-Kass Council on Bioethics are replete with examples of hectoring by articulation of negative possibilities: Human reproductive cloning might be used for eugenics purposes. Reproductive cloning might be an enormous step in the direction of transforming human procreation into human manufacture. The mental and emotional life of a clone might be undesirable. The cloned individual's sense of individuality may be confused by his origin. His connection to others, and particularly to his own family, may become muddled as well. Human cloning might alter the institution of the family, and the way in which individuals and communities come to think of procreation. And so on.

Well, yes, indeed—and nearly every one of these negative possibilities was put forward to oppose in vitro fertilization as a life solution for some infertile couples who wanted children. The basic idea seems always to be the same: New technology is dangerous.

Maybe so, but it's not a sufficient reason for the human species to sit on its haunches and grind flint to start a fire. Even Charles Krauthammer, a strong opponent of all forms of cloning, recognized this in one of his statements to the council: "In a free society, the burden of proof is on those who want to stop [cloning]. So, you do not have to have a good argument to clone; you have got to have a good argument to say why you should not."

So far, apart from articulation of negative possibilities, the good arguments are not in evidence.

CLONING AND POLITICS

THE FINAL REPORT ON CLONING emanating from the Bush-Kass Council on Bioethics constitutes one of the most blatant cases in memory of government manipulation of scientists and science policy. An amazing irony is that the chairman of the commission, a philosopher who calls himself an "ethicist," may have himself possibly breached ethics in order to appease the Bush administration.

The story is a typical political scenario. The Bush-Kass Council on Bioethics that considered human cloning consisted of eighteen members. Stephen L. Carter, a law professor at Yale, did not attend most of the sessions. Of the remaining seventeen members of the council, only four were working scientists, their fields molecular biology, neuroscience, internal medicine, and cell biology.

All members of the council approved a ban on reproductive cloning. Discussions of clinical research cloning, however, produced a different outcome, a definite split, with nine members (including all the scientists) voicing opposition to a ban on clinical research cloning and favoring clinical research cloning with regulation. In April 2002, the nine members, a majority, voted to permit regulated clinical research cloning.

Evidently for Kass, a strong opponent of all forms of cloning, this would not do, and it would certainly disappoint and embarrass the Bush administration. On June 20, 2002, Kass introduced the idea of a moratorium on clinical research cloning, which some members apparently understood to be a pause to allow regulations to be established. The length of the moratorium was not discussed. Nine members, however, remained in favor of regulated clinical research cloning without any moratorium. On June 28, 2002, members received the final policy recommendations and many were shocked to read that by a ten to six majority (one member not reachable and Carter not participating), the council recommended a four-year moratorium on clinical research cloning. Two members (not scientists), apparently in discussion with Kass but not in discussion with others, had changed their positions, the result making the final recommendation agree with the views of Kass and the Bush administration. What went on in secret discussion has never been revealed. The two members who switched their positions at the last minute were social theorist Francis Fukuyama and psychiatrist Paul McHugh.

This sort of back-room manipulation to obtain a politically desired objective is commonplace in government, academia, industry, and anywhere people come together to make policy decisions, but that it's commonplace doesn't make it desirable or ethical. The Council on Bioethics brought people from various professions together to openly discuss and reach an intellectual consensus on specific science policy. To have such manipulations occur inside an ethics commission is a striking example of how politics corrupts reason.

Writing about this affair, the journalist Stephen S. Hall tells the story of a Rockefeller University scientist who kept his research embryos in a freezer for more than two years awaiting a policy decision because he understood that his responsibility to society was greater than his professional objectives. Comparing the behavior of this scientist with the behavior of Kass the ethicist, Hall asks: "Who behaved with greater respect for the community at large, the scientist or the ethicist?"

FUTURE PROSPECTS

THE FOUR-YEAR MORATORIUM FORBIDDING CLINICAL research cloning decreed by the Bush administration during its first term, the policy "recommended" by the Kass Council on Bioethics, expired in the fall of 2005, and anti-cloning forces are apparently preparing to push through a permanent ban of all cloning, both reproductive cloning and clinical research cloning. Leon Kass is evidently firmly committed to banning both types of cloning, but he wants reproductive cloning and clinical research cloning to be considered as two separate issues. Some anti-cloning members of Congress instead prefer a single law prohibiting all cloning. If the recent past is any guide, any input from the scientific community will be ignored. "We have today," Kass has declared, "an administration and a Congress as friendly to human life and human dignity as we are likely to have for many years to come." He suggests that anti-cloning forces respond to the previous inability to pass a total cloning ban in Congress "not by yielding ground but by seizing the initiative."

THE QUESTION POSED AT the head of this chapter—will we have a hundred Jerry Falwells?—can be answered in terms of probabilities: It can't happen now, it probably won't happen during the lifetime of anyone living, but successful cloning of humans will possibly occur during the second half of the twenty-first century, and will probably be a commonplace in the twenty-second century. Science and technology have their own agendas, to understand and control the world, and they don't much care about Luddites in or out of three-piece suits. The question is not if, but when, and if history is any guide, society will adapt to new knowledge the way it always has and move forward into a new phase. We're humans, and that's the way of our species.

PART SIX

Genes, Behavior, and Race

17

GENES AND BEHAVIOR:
YOU NEED A GOLF GENE HERE

Gradual evolution by natural selection might have produced a specific psychological mechanism that generates "pet love" today because in the past such a mechanism promoted healthful feelings of companionship with pets.

— JOHN ALCOCK

Evolutionary psychology wants to have it both ways. It longs after the prestige of hard science but hopes to be held to a lower standard of rigor than, say, molecular biology.

— H. ALLEN ORR

FOR THE PUBLIC, ANY attempt to apply science to the understanding of human behavior constitutes a keg of dynamite waiting to be detonated. It makes no difference whether one is politically left, right, or center, the relationship science + human behavior = politics is the governing equation. And the one area of inquiry that consistently detonates more dynamite than any other is the study of the genetic determinants of behavior, particularly the research areas known as sociobiology and evolutionary psychology.

Only recently has the field of sociobiology broadened its focus to include evolutionary studies of the behavior of social animals more complex than social insects. As for the evolution of human behavior, this intellectual territory is claimed by both evolutionary psychology and cultural anthropology, with the former for the most part a subdivision of sociobiology, emphasizing genetic evolution, and the latter for the most part emphasizing the evolution of culture and the impact of culture on individual behavior.

My personal view is that the academic vitriol of the "sociobiology wars" of the past several decades was not necessary and ought to have been avoided, but

I also believe that if the political left sometimes exaggerated the dangers of sociobiology, the claims of the political right that the concerns raised by the left should be dismissed as irrelevant were counterfeit. The issue was real and remains real—the potential abuse of speculation that wears the cachet of science—and anyone's denial of this issue is a disservice to the public.

SOCIOBIOLOGY

BIOLOGICAL EVOLUTION IS THE EVOLUTION of genes. The field known as "sociobiology" began as the study of the behavior of social insects, especially the behavior of bees, ants, and wasps, and thrived upon the application of evolutionary ideas, especially natural selection, to explain specific insect social behaviors. E. O. Wilson's famous tome, *Sociobiology: The New Synthesis*, published in 1975, is usually considered the starting point for the modern idea that evolutionary biology can be rigorously applied to the study of behavior in animals. Wilson had been studying the behavior of the social insects for years, and no one disputed his claims for the efficacy of evolutionary analyses of such behavior.

But in his 1975 book, Wilson went further and suggested an evolutionary analysis of human behavior was in order, and that many aspects of human behavior could be studied the way animal behavior had been studied—by the application of the idea of genetic evolution via natural selection.

In 1975, the predominant view of Wilson's colleagues, and also the view of many ethologists, researchers who studied "instincts" and behavior in higher animals, was quite different. One such ethologist was Nikolaas Tinbergen (1907–1988), who was awarded the Nobel Prize in Physiology and Medicine in 1973, and who is considered one of the founders of ethology. Although researchers who approach human behavior in terms of genetic evolution often cite Tinbergen as one of their intellectual sources, Tinbergen approached human behavior primarily in terms of the evolution of culture, rather than in terms of the evolution of genes. Here is an excerpt from a lecture given by Tinbergen at Oxford University, October 27, 1964:

> It has often been pointed out that Man, himself a product of evolution of a type similar to that which has created all other animal forms, namely adaptive hereditary change, has now embarked on a new type of evolution, which Hux-

ley calls "psycho-social evolution." I prefer the term "cultural evolution." It is based on accumulated transfer, by tradition, from one generation to the next, of knowledge (or phenotypic) behavior changes, i.e., changes acquired through individual experience. Our culture is very different from that of Cro-Magnon Man, but generally we may not have changed much—most of our modern attributes are due to the accumulation of transferred knowledge. We differ from animals not merely in the extent of what we can ourselves learn, but in the progressive (and steadily accelerating) accumulation of experience through the generations.

This idea, not at all new, of "time-binding" by the human species—the unique transfer of acquired knowledge from one generation to the next—is, I think, the most important difference between the origins of current human behavior and the origins of animal behavior, and is precisely the difference almost totally ignored by modern evolutionary psychology. It's also the primary reason why many biologists, psychologists, and cultural anthropologists rejected—and still reject—current evolutionary psychology as a sufficient strategy for understanding the origins of current human behavior.

A JUNK SCIENCE?

SOCIOBIOLOGY IS THE STUDY OF causes, the evolutionary origins of behavior in animals and humans. Some sociobiologists introduce a philosophical device that differentiates "proximate" research from "ultimate" research. Sociobiologist John Alcock, for example, says:

Studies of how cellular mechanisms and system-operating rules influence behavior are classified as proximate research, which examines the immediate causes of the traits of interest. In contrast, questions about the adaptive (reproductive) value of behaviors are labeled ultimate questions, not because they are more important than proximate ones but because they are different, dealing with the long-term historical causes of the special abilities of the species.

Unfortunately, in writings about human sociobiology, sociobiologists are often slippery: They will admit that culture can be a proximate cause for human behavior, but then they add that since human culture is made possible by genes that determine the structure of the brain and how it works, human genes are involved anyway. Slippery, indeed, and not to the point: The controversy is not whether genes are involved in any way in human behavior, but whether, as E. O. Wilson suggested, genetic evolution via natural selection directly determines cultural proximate causes. Some critics of sociobiology have called this attitude of sociobiologists "intellectual greed"—we can do it all, the sociobiologists say; we don't need the humanities and social sciences to explain anything about human behavior; beauty, love, sex, violence, altruism, art, science, incest, rape, literature—we can do it all; it's all biology, all natural selection, all genes, and if you don't believe that then you're just antievolution and antiscience!

Philip Kitcher likens this attitude of some sociobiologists to shouting with a megaphone: "With great excitement, pronouncements about human nature blare forth." The megaphone, the lack of caution, the insistent refusal to admit contrary evidence has brought human sociobiology a description as a junk science.

BIOLOGICAL EVOLUTION AND CULTURAL EVOLUTION ARE DIFFERENT

WHAT IS IMPORTANT IS THE element of time, since biological evolution requires enormous periods of time before effects are apparent. Consider, for example, human sexual behavior, certainly the most "biological" of human behaviors. In humans in advanced societies, courtship behavior changes rapidly, in ten or fifteen years, or at most in a generation. Courtship behavior, sexual styles, acceptable or unacceptable practices, sexual attitudes—all can undergo severe transformation with great rapidity, and none of these changes can possibly involve the evolution of genes. There is absolutely nothing to suggest that sociobiology or evolutionary psychology, as genetic-evolutionary approaches, have anything to contribute to such questions. The sociobiologist may reply that the current biological context for human sexual behavior has evolved through natural selection of relevant biological processes and structures, i.e., genes, but there is no evidence to support the notion that such genetic evolution has produced any detectable changes during the geological time-blip of the entire history of human civilization, let alone during the past 500 years.

So sociobiologists obsessed with natural selection will not have much to contribute to an understanding of human behavioral changes during a time frame of only centuries or even of millennia. For humans, evolving culture is evolving environment, which means humans are like animals living in a pervasive environment undergoing rapid change, the speed of change blistering compared to the time required for genetic evolution by natural selection. That is a reality, and to the degree that sociobiology twists or hides that reality, sociobiology is a junk science.

EVOLUTIONARY PSYCHOLOGY

SOCIOBIOLOGY FOCUSES ON ANIMAL SOCIETIES, including human societies. Evolutionary psychology, in contrast, focuses on human individual behavior. The proposals by evolutionary psychologists that human behaviors are generally a result of evolutionary natural selection of such behaviors have kicked up intellectual dust storms that have clouded many college campuses for more than thirty years. Whether evolutionary psychology is real science is debatable: The field is far from rigorous, far from evidentiary, and currently it's more a branch of philosophy than a field in science.

As for the applications of evolutionary psychology to public discourse and to public policy via the media, the usual case is that the applications are for the most part junk science and dangerous. Propositions that human behavioral traits are derived through genetic evolution and therefore transmitted via genes are also propositions that human behavioral traits are inherited, and such propositions are welcomed and used by socially destructive political elements such as white supremacists, neo-Nazis, proto-fascists, and proponents of inherited privilege—all of whom adore the idea that behavior is determined by genes. The fears about the new sociobiology that many people had in the 1970s have certainly been realized: Not only is evolutionary psychology a junk science, but its conclusions are politically incendiary and dangerous.

Does this mean the genetic bases of behavior should not be a focus of research? No, it does not, but researchers into inherited human differences who write for the public need to be aware of how their words can be used in politics.

Unfortunately, evolutionary psychologists do not do enough to protect the public from confused interpretations of their ideas. Instead, they often augment the confusion with simplistic popularizations of evolution and psychology that delight the media and cause biologists and classical psychologists to grind their teeth in irritation. Public interest may make academic inquiry interesting and ultimately productive, but outside academe, a bamboozled public is not a good thing at all. The evolutionary psychologist Steven Pinker may plead that his ideas about the genetic evolution by natural selection of male attitudes toward women are not meant as approval of such attitudes, but his ideas are nevertheless picked up and toot-tooted by neoconservative mouth-organs such as *The New Criterion* without a protest by Pinker ever in evidence. Neoconservatives, in fact, have done more than Pinker's academic colleagues to make Pinker, originally a researcher in psycholinguistics, famous as an evolutionary psychologist.

Yes, it's true that there should be no barriers to scientific inquiry, but it's also true that inquiries that are potentially socially destructive need to be rigorous and evidentiary, and evolutionary psychology is currently neither.

GENES AND BEHAVIOR

SO GOES THE CIRCUS: GENES, evolution, behavior, "selfish" genes, selfish behavior, the evolution of greed, altruism, violence, genocide, and even the evolution of rape. The usual research method of sociobiology and evolutionary psychology is at least transparent: Look at any current human behavior, identify something "adaptive" about it, and then tell a story of evolutionary-cum-genetic adaptation. As for evidence, if there isn't any, you can plead that's a problem for future research.

Where the storytelling gets sticky is when evolutionary psychologists attempt to talk about single genes responsible for single behaviors, about genetic explanations of the differences in the behavior of human populations, about the inheritance of behavioral traits as the keynote process in human history. The storytelling gets sticky, the choruses on the right and left are shouting at each other, and our desperado of junk science is prancing his horse on the stage with the ghosts of the hawkers of phrenology and eugenics clapping in the wings.

BUT IS THIS MERELY academia at war with itself? What's all the fuss about? The answer to that question is simplest in street language as some racist fool builds himself an argument for mass discrimination and eventual genocide: If behavior is all in the genes, he thinks, then we may as well stop spending money on social programs designed to change bad behavior because the money is wasted.

And in the background a whisper in his brain: Maybe what we need to do is find out how to identify the bad genes and just get rid of the bad-behaving people, or just stop them from producing more bad-behaving people and just let the present ones die out. Maybe then we'll all be able to go jogging at night in Central Park. Doesn't that make sense? It's common sense that if you have good behavior and bad behavior caused by good genes and bad genes, you want to get rid of the bad genes. Gentlemen, by the grace of almighty God, I stand before you here in Congress to offer an amendment that would once and for all solve the social problems of these United States of America.

Labeling any human behavior as genetic in origin is biological determinism, and we've seen the dangers of that approach in our discussion of the eugenics movements of the first half of the twentieth century, movements that led to sterilization and genocide of people with supposedly "bad genes." When biological determinism gets translated into political policy, it's extremely dangerous.

But how things are labeled does not make a science junk science. The important question is how is the research carried out in evolutionary psychology?

AN EXAMPLE OF RESEARCH IN EVOLUTIONARY PSYCHOLOGY

LET'S CONSIDER A TYPICAL RESEARCH scenario in evolutionary psychology, this one presented by the sociobiologist John Alcock in 2001 in his book *The Triumph of Sociobiology*.

We focus on the hypothesis of evolutionary psychology that males are usually more interested in personal power and status than women because acquisition of personal power and status has had greater positive effects on male than female reproductive success over the evolutionary history of the human species. Thus, goes the argument, natural selection acting on variation in male psychological mechanisms in the past has resulted in the spread of those psychological tendencies likely to promote strong interest in and desire for political influence and high social status.

Alcock then states the hypothesis can be tested. For example, if it is true that at the "ultimate level" striving for high social status is an evolved male reproductive strategy, then in other species in which the trait occurs, not just humans, we expect to see a correlation between male social status and access to sexually receptive females. What one finds is that competition among males for positions of social dominance takes place in many mammals, and when it does, winners typically are rewarded with relatively many mates.

Alcock then proposes that the above generalization applies to our species as well. In traditional preindustrial societies that permit polygamy, dominant males of high social status have more wives on average than those lower in the male pecking order and they have more children. In modern societies that forbid polygamy but allow males to have sexual relationships with more than one woman, rich men have more copulatory opportunities than poor ones.

Finally, Alcock declares: "Thus, sociobiologists have reason to believe that an evolutionary explanation for the intense male drive to compete for high social status has been tested and supported to some extent."

That's it. That's the entire research scenario, hypothesis stated and hypothesis "tested": The human behavior in question is hereditary and an adaptive trait evolved by natural selection of genes.

Is this science or storytelling? Is it any wonder that many scientists call such sociobiology scenarios "fairy tales"? What is the method here? First, identify an apparent human trait, then hypothesize that the trait is a result of Darwinian evolution by natural selection of genes, then look at animal behavior to see if there are analogous traits that produce consequences in apparent agreement with the hypothesis, and then declare the discovery of such in animals is a test (proof) of the hypothesis in humans. But a successful search for animal analogues of human behavior is not a "test" of any hypothesis about genetic evolution by natural selection in humans. No, sir, this is not real science, it's junk science, and any announcement to the public that the apparent desire of men to achieve political power and high social status is an adaptive trait genetically evolved by natural selection—a trait adaptive because the trait increases copulatory opportunities—is a disservice to the public and to science itself.

SOCIOBIOLOGY AS ZOOCENTRISM

THE IMPORTANCE OF SOCIOBIOLOGY'S FOCUS on natural selection must be stressed. More than twenty years ago, Stephen Jay Gould pointed out that sociobiology is not merely any statement that biology, genetics, and evolutionary theory are involved in human behavior. On the contrary, sociobiology is a specific theory about the nature of genetic and evolutionary input into human behavior, a theory based on the idea of the primacy of natural selection as a strategy for understanding human behavior. The theory fragments behavior into "traits," explains traits as adaptive solutions, and claims that each trait is a product of natural selection and rooted in genetic determinants, since natural selection is a theory of genetic change. Gould concluded:

> *Thus, we are presented with unproved and unprovable speculations about the adaptive and genetic basis of specific human behaviors: why some (or all) people are aggressive, xenophobic, religious, acquisitive, or homosexual. Zoocentrism is the primary fallacy of human sociobiology, for this view of human behavior rests on the argument that if the actions of "lower" animals with simple nervous systems arise as genetic products of natural selection, then human behavior should have a similar basis.*

A STORY OF BAD GENES

YES, AS GOULD SAYS, THIS is all about genetic products of natural selection. Consider a story of "bad genes." The early 1970s witnessed a failure of the application of human genetics to human behavior that ought to have made evolutionary psychologists and the proponents of a human sociobiology extremely cautious. In human males, each cell normally has forty-six chromosomes and a so-called XY genotype, one female chromosome (X) and one male chromosome (Y). But a small percentage (less than one percent) of males have an extra Y chromosome and exhibit an XYY genotype. The first human male with an XYY genotype was discovered in 1961. In 1965, P. A. Jacobs et al. reported a link in males between this chromosome abnormality and tall stature, aggressive behavior, and "criminality." In 1966, the case of Richard Speck occupied public attention in America. Speck murdered eight nurses in a Chicago dormitory, and the subsequent trial had the public spellbound. At the Speck trial, his attorney claimed that Speck was innocent due to uncontrollable urges caused by his XYY genotype, but the jury convicted Speck of murder anyway. (Speck was later found to have only a normal XY genotype.)

Genes were suddenly in vogue as an explanation for criminality. The media called the extra Y chromosome the "criminal chromosome." By 1968, researchers at Harvard University were screening all newborn males at Boston Hospital for Women for XYY genotypes and following up the screening by studying the development of the children with the abnormality. The research was funded by a grant from the Centers for Studies of Crime and Delinquency of the National Institute for Mental Health.

A few years later, in 1973, a study by L. F. Jarvik et al. seemed to confirm the relationship of the XYY genotype with criminality with evidence that the frequency of the XYY genotype in the male criminal population was fifteen times that of the normal XY genotype. The Jarvik group proposed that if an extra Y chromosome leads to unusually aggressive behavior, the genes of the single Y chromosome in the normal male could be the origin of violent conduct in normal males.

All of this was happening at the time E. O. Wilson's ideas underwent a conversion from a purely animal sociobiology to a proposal for a human sociobiology. The idea of human behavior determined by biology appealed to enough people to start a movement that soon led to the birth of evolutionary psychology. Unfortunately for the movement, however, the XYY genotype findings as evidence for human biological determinism survived only a few years. In 1976, H. A. Witkin et al. reported that in a study of more than 4,000 Danish men, XYY males tended to commit crimes against property rather than crimes against people, and that XYY males did not exhibit significantly greater evidence of aggressive behavior than did the normal XY control group. XYY males did show a higher average crime rate, but the genetic abnormality was so rare it could not be an explanation for criminality. Of 4,293 criminals analyzed in twenty separate chromosome studies as of 1976, less than 2 percent were found to have the XYY genotype. A consensus in psychiatry declared that any increased criminality in XYY males is not due to specific "criminality" genes, but to apparent lowered intelligence in such males mediating behavior and leading to trouble with the justice system. The criminal chromosome bubble immediately vaporized.

The XYY story is an example of how prone both scientists and the public are to grasp at simple explanations for complex phenomena. Human behavior is about as complex as things can get in biology, and because of this we need to be especially cautious in the search for biological determinants. Are evolutionary psychologists cautious enough?

EINSTEIN'S BRAIN

IN 1999, IN A *NEW York Times* OpEd-page commentary on a recently completed study of Einstein's brain by brain histologists, evolutionary psychologist Stephen Pinker wrote:

> *The elegant study . . . is consistent with the themes of modern cognitive neu-*
> *roscience. Every aspect of thought and emotion is rooted in brain structure*
> *and function, including many psychological disorders and, presumably, ge-*
> *nius. The study confirms that the brain is a modular system comprising mul-*
> *tiple intelligences, mostly nonverbal.*

But although one way to view the brain (Pinker's way) may indeed be as a "modular system," the histological study referred to by Pinker confirmed nothing about the brain except some apparently unusual details in a small region, the details summarized by the authors of the study as follows:

> *The gross anatomy of Einstein's brain was within normal limits with the ex-*
> *ception of his parietal lobes. In each hemisphere, morphology of the Sylvian*
> *fissure was unique compared with 182 hemispheres from the 35 control male*
> *and 56 female brains: The posterior end of the Sylvian fissure had a rela-*
> *tively anterior position, associated with no parietal operculum.*

Small differences between people of gross folds and fissures of the brain are not unusual, and there is certainly no basis for relating any such variation in Einstein's brain to his talents.

The term "histology" refers to the structure of tissues as revealed by microscopy and not to gross anatomy as revealed by the naked eye. The study found no significant histological differences between Einstein's brain and any other average brain. Important histological or cytological differences might be there, but they have not yet been identified.

No particular view of the brain was "confirmed," certainly not a view of the brain as a "modular system." The authors of the study make no pretense about implications when they say: "This report clearly does not resolve the long-standing issue of the neuroanatomical substrate of intelligence."

The title of Pinker's commentary was "His Brain Measured Up: Einstein Still Solving Scientific Mysteries." Unfortunately, not any more. No mysteries solved here. There is no evidence that the small gross morphological differences between Einstein's brain and the ninety-one control brains are necessary to produce Einstein's talents, and no evidence that these differences are not present in people who have no talents at all. We lack an evidentiary science relating macroscopic and microscopic brain morphology to mental abilities. We may get there in the future, but we aren't there yet.

Meanwhile, this sort of pop science exaggeration in the media by someone with an authoritative academic title only adds to the public confusion concerning the relation between genetically evolved brain structure and culturally evolved behavior. In his most recent book, *The Blank Slate* (2002), Pinker's comment about Einstein's brain is muted: "A study of Albert Einstein's brain revealed that he had large, unusually shaped inferior parietal lobules, which participate in spatial reasoning and intuitions about number."

But on the same page Pinker says, "According to a recent study of the brains of identical and fraternal twins, differences in the amount of gray matter in the frontal lobes are not only genetically influenced but are significantly correlated with differences in intelligence."

Wonderful. Now we have a serious problem, since the histological analysis, cited by Pinker, that found a small difference between Einstein's parietal lobes and the parietal lobes of other brains also found absolutely no differences between Einstein's frontal lobes and the frontal lobes of other brains. Are we then to conclude, following Pinker's ideas about relating gross brain structure to mental abilities, that Einstein was a man of only average intelligence but with a talent for "spatial reasoning and intuitions about number"?

Since he's an evolutionary psychologist, Pinker has a view of human behavior that demands a biological substrate evolving by genetic evolution. That's fine, and one wishes him all speed in research. But so far no genetically evolving biological substrate has been found that distinguishes Einstein's brain in any important way, and a mere public announcement that such a substrate is present in Einstein's brain is not good enough. Maybe Einstein's genius was due to a sudden environmental hormonal phenomenon that occurred during fetal development or during his fifth postnatal year. Do we know? Has there been any demonstration of any other cause? Why use Einstein to "confirm" genetic evolution of brain and mind by natural selection? The public may be entertained by storytelling about Einstein's brain, but such storytelling is not real science, and the public is being misinformed about what we know concerning the relation between biology and behavior.

SOCIOBIOLOGY AND BIOLOGICAL DETERMINISM

SERIOUS SOCIOBIOLOGISTS NEED TO DO more to dispel popular junk-science interpretations of their work, particularly interpretations by the media and by politicians. Sociobiologists often complain that they are falsely labeled as genetic determinists, but when their popular writings are interpreted by political conservatives as support for genetic determinism, sociobiologists rarely protest. The result is that sociobiologists who do research into the genetic factors related to complex human social problems are seen by their academic opponents as diverting attention from and ultimately undermining attempts to ameliorate the socioeconomic conditions related to those social problems. In the view of these opponents, genetic differences are likely to be less important than social inequalities in determining most human behavior. Furthermore, they argue that the end result of a biological determinist perspective (especially a perspective hardly based on evidence) is discrimination against, rather than help for, those who are deemed inferior or defective. If the history of political genetics in the United States in the period 1920 to 1970 is any guide, this is exactly what will happen if and when genetic determinism is ascendant. If sociobiologists are truly opposed to such an outcome, it would help if they more actively opposed media distortions that lead the public to believe myths about genes and the inheritance of human behavioral traits.

GENETICS, HUMAN DISEASE, AND GENETIC TESTING

OUTSIDE THE CONTROVERSIES CONCERNING GENETIC evolution and behavior, the involvement of genetics in human disease has become a subject of importance to both medicine and politics. When mutations in specific genes are related to inherited human disease, there are good clinical reasons to identify those mutated genes in order to make various treatments possible, and to design tests to screen for those genes in patients who are willing to be screened. There will come a time when the governments of all advanced nations will need to deal with the problems of commercialization of genetic testing and pressures by insurance companies and employers that force individuals to be screened for various gene mutations suspected to be involved in various medical pathologies, including the diagnostic zoo of psychiatric pathologies. There will be much abuse, and a good deal of that abuse will involve twisted realities presented to the public.

As early as 1995, over fifty biotechnology companies were developing or providing tests to diagnose genetic disorders or to predict the risk of their future occurrence. Common complex disorders, usually disorders of adult onset such as Alzheimer's disease and breast and colon cancer, make up the single largest category for which tests are under commercial development. These companies often exaggerate claims for predictive tests for common complex disorders. The exaggerations include claims for clinical validity, claims about the probability of a detectable susceptibility-conferring gene occurring in those who would get the disease, and claims about the probability that those with a susceptibility-conferring gene would actually get the disease, and also claims about how a positive test result could help people cope with future disease.

In most cases, tests for mutations in one or more genes will be tests for "susceptibility" genes, which means a positive test result will identify only a probability that an individual will at some time in the future be diagnosed with some specific clinical pathology. Without public education, this will become a fertile field for junk-science manipulation of social agendas. The identification of people with such probabilistic genetically based pathologies is of great value to commercial health insurers and other interests, and there is thus a danger of branding of potential commercial insurance clients, employees, children in various social contexts, and others on the basis of genetic testing for "susceptibility genes." A special problem will involve genetic testing for psychiatric susceptibilities, since psychiatric diagnostics is a continuing mess of changing criteria and shifting uncertainties. The "criminal chromosome" fiasco of the early 1970s may be history now, but it's certainly a warning about troubles that lie ahead of us.

The current consensus in the medical community is that genetic screening is only appropriate when the biological nature of the disease is understood; the screening tests are valid and reliable; the sensitivity, specificity, false-negative and false-positive rates are acceptable; and effective therapy is available. How likely is it that these criteria, under pressures from government and industry, will be maintained in the future?

BECAUSE THE PUBLIC UNDERSTANDABLY has an intense interest in fathoming human nature, the public quickly grasps anything brought to its attention that carries the aura of "science." Human behavior is complex, and it depends on individual development, experience, family environment, social environment, culture, and the constraints of evolved physiology. H. Allen Orr, in the quotation at the head of this chapter, points out the lack of rigor in evolutionary psychology. I think Orr is on the mark about this, but I don't know if the defect will be permanent. What I do believe is that at the present time evolutionary psychology is handicapped by biologists who are only amateur psychologists and psychologists who are only amateur biologists, and the result has been a great deal of dust kicked up everywhere. How long it will take for the dust to settle remains uncertain.

Meanwhile, since this century will undoubtedly be a century of the human genome, the future will too often bring us junk-science storytelling announcements of discoveries relating genes to behavior. Caveat emptor.

18

RACE AND IQ: TWO MYTHS MAKE A BUMBLE

> Evidence from a number of independent studies of racial differences in cranial capacity and in autopsied brain volume and brain weight show approximately 100 grams difference [black brains weighing less than white brains] between the average autopsied brain weights of blacks and whites matched for sex, age, and overall body size.
> —ARTHUR R. JENSEN, LEADING ADVOCATE OF A GENETIC BASIS OF RACIAL DIFFERENCES IN "GENERAL INTELLIGENCE," 2002

> Autopsy results: Weight of Albert Einstein's autopsied brain: 1230 grams. Average weight of 91 controls ages 30–70 years: 1400 grams. Average weight of 8 age-matched controls aged 65 years: 1386 grams.
> —B. F. WITELSON ET AL. IN THE MEDICAL JOURNAL *THE LANCET*, 1999

AMERICA LIKES TO PRESENT itself to the world as the archetypical "melting pot" of races, ethnicities, nationalities, and so on, but no country that exists or ever existed has been more obsessed with describing, detailing, measuring, and analyzing group differences, most often to rank groups in terms of abilities, general or special, and particularly in terms of the ability called "general intelligence." And the professional cadre most involved in this obsession has been the band of psychologists called "psychometricians," the test designers, and test administrators, and test analyzers who have been on constant duty in education, government, and industry with their test calipers at the ready. We melt together in America, only to be filtered out and repoured into separate and separating molds so convenient for discrimination. How many people are aware that the ubiquitous Scholastic Achievement Test (SAT) was originally invented to control the entry of Jews into higher education?

The term "race" generally refers to a human population distinguished as a more or less distinct group by genetically transmitted physical characteristics. But which physical characteristics? And if the question which physical characteristics is a reasonable question, what is the implication for any attempt to relate a specific set of physical characteristics to behavior—in particular, to something as nebulous as "general intelligence"? In Nazi Germany in the 1930s, a group of prominent anthropologists devoted their professional careers to investigating the correlations between what they considered to be the physical characteristics of the Jewish "race" and what they considered to be the "inferior" traits of Jews. The Nazis were out of power only a few decades when in America a group of prominent psychologists led by A. R. Jensen began investigating the correlations between "race" and "general intelligence" in American subpopulations. Through the remainder of the twentieth century, these psychologists fed the American public the idea that Asians and Jews are more "intelligent" than non-Jewish whites, who in turn are more "intelligent" than blacks, and the idea that these differences are inherited and unchangeable. When, in 1969, Jensen was asked how much IQ and scholastic achievement could be boosted in blacks, his answer was, "Not much." That is still his answer today, thirty-seven years after publishing the article in the *Harvard Educational Review* that brought him instant fame.

Questions immediately suggest themselves: Is "general intelligence" a real human trait? Are the research measurements valid? Are the conclusions about heritability and unchangeability valid? Are the "racial" categories used in this research meaningful?

One thing is clear: If this subspecialty of individual- and group-differences research in psychology is junk science, it's junk science of great potential danger to the structure of society—and the public, policymakers, and scientists in every field of science need to be aware of what has been happening since Arthur R. Jensen and his followers, in 1969, took up the mantle of providing "scientific evidence" for inherited racial differences in intelligence.

Is it junk science? My answer is yes.

THE G FACTOR

JENSEN'S RESEARCH HAS INVOLVED WHAT is termed the "g factor," an entity already proposed in 1904 by the British psychologist Charles Spearman (1863–1945). The g factor is an operationally defined measure of what is considered to be "general intelligence," and intense debate has raged about what it is or is not. Jensen claims that "Spearman's g is as important to psychology as Newton's law of gravitation is to physics." And further: "The g factor is not the result of some mathematical manipulations. There is no longer any doubt of the physical reality of g."

Jensen has always considered g to be of monumental importance. By 1979, Jensen's view of the g factor had flowered into a remarkably grandiose notion of g as a biological entity:

> The common features of experimental tests developed by comparative psychologists that most clearly distinguish, say, chickens from dogs, dogs from monkeys, and monkeys from chimpanzees suggests that they are roughly scalable along a g dimension. . . . G can be viewed as an interspecies concept with a broad biological base culminating in the primates.

So what is this marvelous g factor?

The g factor is an entity that depends on a method called "factor analysis," first developed in the early part of the twentieth century by statisticians and psychologists in Edinburgh, London, and Chicago. Factor analysis is essentially a method for analyzing intercorrelated data sets into independently variable "factors," with each factor defined by the degree to which it participates in each data set. The most important factor is produced first, and if one continues the analysis, one can derive lesser factors, then residual factors, and so on. One does not need to be a mathematician to use factor analysis, any more than one needs to be a mathematician to use algebra. Fifty years ago, when I used factor analysis in graduate school to the point of nausea, the work was arduous because we had no computers. These days it's a snap, a routine analysis, provided you have both data and questions appropriate for the method.

Consider the following situation. We have ten computers lined up on a long table. They vary in hardware, operating systems, and software, but we know nothing about the details of any of this variation, nothing about what is inside these machines. All we can see are the keyboards, the processing towers, and the screens of the monitors. Our problem is to study the comparative performance of these machines on various tasks. We design a set of problems (inputs) for these machines to solve, problems whose solutions can be objectively scored on some scale, and we systematically present our designed problems to these machines to obtain and score their respective solutions (responses). We find that for any specific machine, scores for various problems are positively correlated—if a machine does well on one sort of problem, it will usually do well on another sort of problem. We subject all sets of scores to statistical analysis of a particular kind (factor analysis), and we find we can extract a single general factor, which we call the "performance factor," that accounts for the variance of scores of the set of machines. We can therefore rank each machine in accordance with its numerical performance factor, although we have no idea what this factor actually relates to in terms of differences in hardware, operating systems, or software. The performance factor, or "p factor," for each machine is an operationally defined parameter that may or may not be useful, depending on what one wants to know about these machines and about the differences between them.

The Spearman-Jensen "g factor" is essentially equivalent to our performance factor, and although some have questioned whether "g" is "real" or an artifact, the g factor remains operationally defined and its measurement reliable from one set of observations to the next. It may not represent anything useful, but anyone who exactly repeats the procedures used in determining "g" will obtain the same value of "g" as anyone else who repeats the identical routine.

However, I do dispute Jensen's claim that "g" has a "physical" reality. In Jensen's own words, "a factor is defined as a latent variable, or a hypothetical source of variance, that is common to two or more variables." Indeed, and in this case the "hypothetical source of variance," the g factor, has never been related in any convincing way to any physical entity or process except the trivial one of the whole brain itself as the source of performance. The g factor has something to do with the human brain, but what and how remain unknown.

Stephen Jay Gould believed that "the key error of factor analysis lies in reification, or the conversion of abstractions into putative real entities." But I think I must disagree with this, too, since factor analysis is merely a method of analysis that itself requires no conversions. The reification error is made not by the method but by the psychometricians who use factor analysis and then make conversions from mathematical abstractions to real entities.

What the g factor does have is not a "physical reality" but an operational reality, and no more than that. The difference can be important. A trivial illustration is that the average weight of a group of people in a room has an operational reality but not a physical reality—it's a construction. Given only the average weight, one cannot make any statement at all about any individual in the room. The temperature of a gas is also a construction, an operationally defined indicator of the direction of heat flow with respect to another system. Given only the temperature of a gas, one cannot make any statement at all about any of the molecules of the gas, or the chemical nature of the gas, or whether the gas has just exploded or will explode in the next second.

The g factor is an operationally defined construction that by itself allows no statements about processes in the nervous system that produced the performances that were originally measured and analyzed to produce the g factor.

So although some people have done battle with the "reality" of the g factor, I don't think such a battle is useful. What Jensen and others do with the g factor, however, is something else: They claim the g factor, "general intelligence," is for the most part inherited and therefore unchangeable. Here I indeed do battle with Jensen and his followers and with what I think is their unnecessary neglect of the fundamentals of biology.

HEREDITY AND INTELLIGENCE

JENSEN'S STATEMENTS ABOUT THE HERITABILITY of general intelligence are unambiguous:

> There is no valid environmental explanation . . . for why identical twins
> reared apart should be almost as alike [in general intelligence] as identical
> twins reared together and much more alike than fraternal twins or ordinary
> siblings reared together; or, on the other hand, why unrelated adopted chil-
> dren reared together should be so unlike.

And again:

> In all the published studies of pairs of monozygotic (identical) twins sepa-
> rated in infancy, put out for adoption, and reared entirely apart, the average
> correlation between the IQs of the twins in adulthood is 0.75. Since the
> twins did not share a common environment but do have identical genes, this

correlation provides a fair estimate of the heritability of IQ for people in the
normal range of environments for our population.

Jensen's argument that the g factor is for the most part inherited is based on twin studies, particularly studies of monozygotic twins reared apart. In 1969, when Jensen published his sensational article (after making certain a reporter for the then conservative news magazine U.S. *News and World Report* received a prepublication copy of the manuscript), he based his argument mainly on the twin studies of the British psychologist Cyril Burt. When Burt's twin studies were subsequently trashed by nearly every psychologist in the world as fraudulent, Jensen removed Burt's twins from analysis and substituted twins from other studies. But which twins are used for analysis is not the issue: Both of the statements by Jensen quoted above involve an important fallacy that Jensen has never dealt with adequately and that he still refuses to admit casts doubt on his conclusions.

Let me state the fallacy concisely.

If one recognizes that human "intelligence" depends on activity of the human nervous system and perforce to some degree on the way the nervous system develops, then it must also be recognized that the most important "environment" for the developing nervous system is not, as Jensen believes, the postnatal childhood environment, but instead the prenatal fetal environment from conception to birth, the period during which the human nervous system undergoes its most profound architectural development, i.e., its "hardwiring." Rapid histological and cytological development of the human brain cortex also occurs in the early years after birth, but none of the postnatal changes are as profound as the prenatal changes. It's lack of consideration of the fetal environment that turns the conclusions of Jensen and his followers into mere speculations: Their "evidence" consists of results of studies without adequate controls.

Concerning identical twins and fetal environment, it's important to note that identical twins that share the same uterus do not always share chorionic and placental systems, and that common differences in birth weight of identical twins provide prima facie evidence that identical twins need not have identical fetal environments.

Is there any explanation for the lack of focus on what is obviously an important variable in neural development, the fetal environment? In the 1960s, most psychologists, including Jensen, were "nonbiological"—they lacked training in the biological sciences. In addition, most psychologists at that time (and also the public) were focused on postnatal environment, especially home environment and culture, as determinants of intelligence and behavior in general. Fetal environment was not (and has never been) considered an important variable in studies of the origins of human behavior, especially since most kinds of experiments on human fetal environment are currently not possible.

What is important is that studies of IQ performance that do not provide matched controls for fetal environment can tell us nothing about the relative contributions of heredity and environment to the g factor or to any other neurological performance measure. Since studies with such matched controls are currently not possible, there are no available twin studies that provide evidence one way or the other in this matter. Jensen occasionally mentions the possible involvement of what he calls "biological microenvironments" such as the fetal environment and various perinatal conditions, but he never considers these as important variables, always advising us that the effects are "small." Since there are no controlled studies in this area, on what basis are the effects said to be "small"? All the Jensen twin studies are without adequate controls for fetal environment and are thus not acceptable as evidence to support his conclusions about the heritability of the g factor.

In his writings, Jensen mentions again and again that his work is pure science and nothing else. But does he apply the same standards of rigor to his own work as is generally applied in biology, chemistry, and physics? In any research in science, a lack of adequate controls makes conclusions arbitrary, and the drumbeating to the public of such arbitrary conclusions makes that research junk science.

THE IMPORTANCE OF FETAL ENVIRONMENT

HOW IMPORTANT IS FETAL ENVIRONMENT for the human developing nervous system? Suggestions are provided by clinical studies. But first we need to recognize something about "thresholds." The idea that thresholds for "wellness" exist in the fetus and neonate, thresholds such that on one side of the threshold the fetus or neonate is healthy or "normal," while on the other side of the threshold the fetus of neonate is unhealthy—this idea is a fiction, any such threshold merely a convenience for clinical diagnosis. Our general understanding of biological processes suggests caution about thresholds; it's more likely that variation of functions in fetuses and neonates exists along a smooth continuum, with blatant dysfunction easily recognized, subtle dysfunction hardly recognized, and subtlest dysfunction and variation never recognized early. The most easily recognized neonatal dysfunctions are those that produce a dead neonate, but behind neonatal mortality is a continuum of dysfunctions and variations down to the subtlest that are not yet ever clinically or experimentally evident, but which allow an apparent "normal" birth.

This continuum of dysfunctions and variation reflects the continuum of genetic and fetal environmental differences and the interplay between them that produces the neonate dead or alive. The fact of relevance in this context is that neonatal mortality (defined as death at age less than twenty-eight days) accounts for approximately two-thirds of infant deaths in the United States, with blacks continuing to have the highest neonatal mortality rate, a rate more than twice that of any other racial/ethnic population. Biology tells us that behind this statistic must be a continuum of dysfunctions and variation that merges into a continuum of differences in genes and fetal environments. In short, how much of the differences in fetal damage are never recognized? How much of simple fetal environment–driven neurological variability is never recognized? How much does human intelligence depend on fetal environment rather than on postnatal environment? At the present time, we don't have answers to these questions.

And if such differences exist between black and white populations, do they also exist within populations? The idea that in a study of intelligence if one matches people for household income one is also a matching for fetal environments is an idea both simplistic and foolish. Identical household income does not mean fetal environments are identical. Indeed, within the same family, fetal environments of siblings can be different.

Fetal environments remain the crucial uncontrolled variables in studies of the heritability of human traits. We just don't know the details of these environments, and we don't know what these environments contribute to later behavior.

HOW SUBTLE ARE THE differences in fetal environments? A suggested answer is provided by a look at the clinical entity called "fetal alcohol syndrome," a clinical syndrome caused by maternal alcohol use during pregnancy. It's one of the leading causes of preventable birth defects and developmental disabilities in the United States. The syndrome is diagnosed on the basis of a combination of apparent effects: growth deficiency (pre- or postnatal), central nervous system dysfunction, facial dysmorphology, and maternal alcohol use during pregnancy. Estimates of the prevalence vary from 0.2 to 1.0 per 1000 live-born infants. Differences have been noted among racial/ethnic populations, and data indicate that children born to mothers in certain racial/ethnic populations have consistently higher prevalence rates. Findings confirm higher prevalence rates among black and American Indian/Alaska Native populations.

It's also believed that the number of children affected adversely by in-utero exposure to alcohol is probably underestimated. What is important in this context is that there is no apparent basis to assign a threshold: We don't know enough about the pathologies involved for us to say alcohol intake per day below a certain threshold will have no effect at all on the fetus. What is evident is that blood alcohol levels apparently nontoxic for the mother can be very toxic for the developing fetus, and the same is true of many other environmental chemicals that may be nontoxic to adults but deleterious in trace quantities to the fetus (for example, lead and other heavy metals). So not only do we need to admit the fetal environment as a variable affecting development of the fetal nervous system, but we also need to admit that subtle interactions may occur between the fetal environment and development that are beyond our present knowledge. We are not yet able to do controlled research to determine the heritability of any performance function of the human central nervous system.

We also see how maternal environment, the environment of the mother, may impact fetal environment, the environment of the fetus, in studies of tobacco use. During most of the twentieth century, a large percentage of American women smoked cigarettes during pregnancy. Maternal smoking during pregnancy has many identified clinical consequences both during and after pregnancy, such as blood coagulation problems, obstetric accidents, and intrauterine growth retardation. The presence of tobacco-specific metabolites has been described in fetal blood and cell-free amniotic fluid, and in newborns from women who smoke. Recent studies demonstrate that smoking ten or more cigarettes per day for at least ten years and during pregnancy is associated with increased chromosomal instability in fetal amniocytes (cells of fetal origin in the amniotic fluid). But the criterion of ten or more cigarettes per day for at least ten years is arbitrary: One cigarette a day during pregnancy may produce effects that we cannot detect simply because we don't know how to detect them, but effects that can be important for the later performance of the nervous system.

At the present time, the fetal environment constitutes a universe of environmental variables almost totally unexplored by intelligence-heritability research, and the public needs to recognize that the conclusions of Jensen and his followers from their studies have been biologically simplistic.

IS INTELLIGENCE UNCHANGEABLE?

FROM THE BEGINNING OF THEIR work, Jensen and his followers have emphasized that not only are inherited traits such as general intelligence unchangeable, but that the apparent unchangeability of general intelligence in actual educational practice is itself evidence of heritability of the trait. The Jensen group is ignoring biological reality: What is unchangeable, except by early general mutation or later local mutation, is the genome, and not the expression of the various genes of the genome. In the past several decades, research in epigenetics, the study of the variables that control gene expression in all its forms, has demonstrated just how changeable the heritability of a trait may be when the relevant expression variables are identified and manipulated. There is no scientific evidence that neurological performance associated with "general intelligence" of an individual or of a population is unchangeable, and all the biological evidence for the importance of fetal environment and the changeability of gene expression suggests the contrary.

INTELLIGENCE AND RACE

IF THE G FACTOR STUDIES of the heritability of general intelligence are invalid because of lack of adequate controls, what does that imply for studies of g factor differences between blacks and whites? One can certainly study differences, but at the present time there is no scientific basis for assigning heritability quotients to such differences—not until adequate controls for differences in fetal environments are in place in research. But beyond that, it's remarkable that Jensen and the group of psychometricians who follow him, and who prattle about how their methods are approaching the rigor of the methods of physics, so cavalierly dismiss the consensus view of the sister science of anthropology that the public categorizations of race, the categorizations made by these psychometricians, have little scientific basis or utility. Has the Jensen group created a new branch of anthropology?

Systematists in anthropology have not defined a "type specimen" for humans, in contrast to the type specimens defined for other species. As for subcategories, or "races," of humans, the racial labels used by the public today (and by Jensen and his followers) are at least as old as the subcategories of the Swedish botanist Carolus Linnaeus (1707–1778). Linnaeus recognized four principal geographic varieties or subspecies of humans: Americanus, Europaeus, Asiaticus, and Afer (Africans). He defined two other categories: Monstrosus, mostly hairy men with tails and other fanciful creatures, but also including some existing groups such as Patagonians; and Ferus, or "wild boys," thought to be raised by animals, but actually retarded or mentally ill children that had been abandoned by their parents.

In the Blumenbach's scheme of 1795, Johann Blumenbach (1752–1840) added a fifth category, Malay, including Polynesians, Melanesians, and Australians. Blumenbach is also responsible for using the term "Caucasian" to refer in general to Europeans. He apparently chose the term "Caucasian" on the basis of physical appearance: he thought Europeans had the greatest physical beauty of all humans, and among Europeans he thought those from around the Caucasus Mountains the most beautiful. Not quite an objective method of classification. More recently, the American Nobel laureate physicist William Shockley, apparently a wannabe anthropologist, suggested simple categorizations by skin color as sufficient. How someone trained in physics can be so cavalierly obtuse outside physics is a psychiatric puzzle, but of course Shockley has had well-known predecessors, and maybe he was merely an unwitting demonstration of the theory of "multiple intelligences."

The extent to which racial classifications of humans reflect any underlying biological reality is highly controversial among anthropologists and biologists, and proponents of racial classification schemes have been unable to agree on the number of races (proposals range from three to more than a hundred), let alone how specific populations should be classified.

In general, biologists and anthropologists, when confronted with the existing degree of variability in human races, conclude that any concept of a "pure" race is fictional. However, on the basis of a common index of "genetic distance," five "races" can be recognized: African, Caucasian, Greater Asian, Amerindian, and Australoid. But members of any of these races are not genetically pure in the sense of sharing a uniform genetic identity, nor does genetic uniformity even apply to members of the same family. As the population geneticist Richard Lewontin has pointed out, more than 94 percent of the genetic variability among humans comes from differences among individuals and groups of the same race, and only 6 percent from differences among races.

JENSEN AND THE PIONEER FUND

THE SO-CALLED GENETIC FALLACY HAS nothing to do with genes or genetic determinism, but is the name given to any violation of the idea that the origin of a proposition should not count in weighing its truth or falsity. So, for example, if the Nazi storm trooper Pascual Jordan made a contribution to quantum mechanics (which he did), that contribution must be evaluated independently of his personal life, and to do otherwise is to commit the genetic fallacy.

I agree. As far as science is concerned, there is no alternative: The truth or falsity of a proposition cannot depend on whether the proposition comes to us from the sky or from under a rock.

But history is another game, and in history knowledge of the origins of an idea is often of historical importance, often of intrinsic "human interest," often knowledge that satisfies human curiosity. So, since I'm a human being, I have my human curiosities, and one curiosity has been the funding for Jensen's research, and the apparent connection to William Shockley, the racist physicist whom we met in the first chapter of this book and again in this chapter, the Nobel laureate who offered "$1,000 a point" to induce low-IQ people to be sterilized: Individuals would receive $1,000 for every point of IQ below 100. Is your IQ 80? You receive $20,000 if you allow us to sterilize you.

It seems the two men, Jensen and Shockley, were connected via the Pioneer Fund, a philanthropic organization apparently created in 1937 to promote white supremacy and racial eugenics. William Shockley evidently recruited Arthur Jensen to the Pioneer Fund, and Jensen has been supported by the Pioneer Fund since the mid-1960s. When recently questioned about his association with the Pioneer Fund, Jensen replied: "The notion that the Pioneer Fund has had anything at all to do with race conflict in the United States or anywhere else is utterly ridiculous as well as totally false."

Let's not go there. It's politics and not science, and I'd rather have others provide an appropriate sociopolitical analysis of this aspect of Jensen's research. The textile magnate and eugenicist Wickliffe Draper, who founded the Pioneer Fund in 1937, had a right to use his money to support his desire to "protect" the

American gene pool; and we, in turn, have a right to ignore the junk science that wriggled in his head.

One never knows what works in the minds of people, but on the surface, at least, Jensen and American white supremacists have always had a common interest, and one not hidden by Jensen at all. Here is Jensen in his famous 1969 *Harvard Educational Review* article: "There are intelligence genes, which are found in populations in different proportions. . . . The number of intelligence genes seems lower, overall, in the black population than in the white."

From a scientific standpoint, these are remarkable statements. Since no "intelligence gene" has ever been identified, now or then, the statement that such genes "are found" is an untruth. As for the number, since no such genes have been identified, how does one count the number of such genes to determine that the number is less in the black population than in the white population? This is not real science. Such statements may have tickled the fancies of journalists and politicians during the past three decades, but it's junk science blatant and unequivocal.

THE FAILURE OF JENSENISM

ALTHOUGH SOME NEW SEPARATIST FANTASY may soon replace it, Jensenism is apparently passing. Concerning the courses he taught for decades at the University of California at Berkeley on behavior genetics and on theories of intelligence, Jensen complains, "It says something about the social sciences in contemporary America that after my retirement no one was hired to teach these courses in either the psychology or the education departments."

About the growing evidence that the heritability of general intelligence is much less than he proposes, Jensen says:"Even if the heritability of IQ is 0.50 rather than 0.78 . . . it still shows that there is a significant genetic component to individual differences in mental ability."

But there is no evidence one way or the other. To illustrate the absence of any evidentiary foundation for Jensen's numbers, I propose the following scheme: variance due to genetics, zero; variance due to fetal environment, 0.70; variance due to culture-home environment, 0.30. My numbers are guesses, but there is nothing in Jensen's data to demonstrate my numbers are not correct, nothing in his data to refute my guess that 100 percent of the variance is due to environment. In the absence of controls for fetal environment,

Jensen has no basis for assigning heritability quotients of 0.78 or 0.50 or any other number, and certainly no basis for thirty years of inflammatory statements that differences between blacks and whites in IQ scores are unchangeable.

The public has been ill-served. But while studies of the genetic bases of neurological performance will continue, the era of Jensenist junk science is nearly finished. In another generation, the g factor, a useless concept in both biology and psychology, will have vanished as a rallying point for people obsessed with racial differences. But don't expect these same people to become beneficent and turn to emphasizing racial similarities. We're not there yet. The divide between people who focus on racial differences and people who focus on racial similarities may be an enduring psychiatric problem in America and elsewhere.

OTHER VIEWS ABOUT HEREDITY AND INTELLIGENCE

I HAVEN'T DEALT HERE SPECIFICALLY with the views of Richard Herrnstein and Charles Murray as expressed in their book, *The Bell Curve*. The reason is I think they don't add much to Jensenism. If Jensenism is junk science, the views of Herrnstein and Murray concerning the heritability of "general intelligence" have no basis and become irrelevant. But certainly Herrnstein and Murray are more cautious than Jensen, and they seem to recognize they are standing on a shaky scaffold when they say, "The nonexperimental social sciences cannot demonstrate unequivocal causality." Yes, that's precisely the point about Herrnstein and Murray and Jensen: No adequate controls, no adequate experiments, no unequivocal demonstration of causality, yet much drumbeating about "evidence" and "solid conclusions"—in other words, junk science. Is it surprising that many critics in the basic sciences respond with irritation that we have no basis yet for conclusions about the sources of individual and group differences in human performance? The Herrnstein and Murray dystopian vision of a future America as a "custodial state" burdened with large numbers of genetically inferior people who will need to be maintained on "a high-tech and more lavish version of the Indian reservation" is science fiction rather than a contribution to science—and socially destructive.

A similar exercise in junk science concerning race and intelligence recently appeared in a book by Vincent Sarich and Frank Miele, a compendium of errors, exaggerations, and speculations about race presented as fact. Sarich and Miele, at least are unequivocal:

There is a significant genetic component in intelligence. It is generally
agreed that the heritability of intelligence within the white population of
Europe, the United States, and Australia is between 0.50 and 0.87 (that is,
between 50 and 87 percent of the individual differences in intelligence in
those groups is attributable to genetic differences), depending on the specific
model used.

In the first place, there is more controversy than general agreement about the heritability of intelligence, and in the second place, as I've demonstrated, there is no hard science behind any current numerical estimate of the heritability of intelligence. Miele is the same journalist who recently published, with Arthur Jensen, an exposition of Jensen's ideas. Vincent Sarich is a physical anthropologist whose controversial ideas about racial categorizations have been dismissed by many anthropologists. The authors use *The Bell Curve* as their primary source for data about the genetic basis of intelligence, but ignore the Herrnstein-Murray caveat that "the nonexperimental social sciences cannot demonstrate unequivocal causality." A statement assigning a number to the heritability of intelligence is a statement about unequivocal causality, and at present such a statement is a fallacy.

What is apparent is that if Jensenism per se is passing, others are willing to carry forward the junk-science banner of exaggerated racial differences.

CRITIQUES OF INTELLIGENCE TESTING

SINCE THE FOCUS OF THIS book is only junk science, I also have not dealt specifically with other views of intelligence. You must not think that Jensen and his cohorts present the views of psychologists in general, since that's not true at all. Most contemporary psychologists do not believe that the concept of "intelligence" is either unitary or useful, and the consensus among psychologists is that expressed concisely by Howard Gardner:

[IQ] tests have predictive power for success in schooling, but relatively little
predictive power outside the school context, especially when more potent fac-
tors like social and economic background have been taken into account. Too
much of a hullabaloo over the possible heritability of IQ has been sustained
over the past few decades, and while few authorities would go so far as to

claim that the IQ is in no degree inherited, extreme claims on heritability within and across races have been discredited.

The late Stephen Jay Gould, in his book *The Mismeasure of Man*, put the concept of general intelligence and the ideas of Arthur Jensen and his cohorts on the chopping block and went to work with his cleaver, causing pandemonium and then silence in that part of the conservative community dedicated to inherited privilege by reason of assumed inherited merit. But do expect the silence to be temporary: Too many people like the idea that their high place in society is the deserved end of good breeding. It makes them feel good. It gives them a sense of satisfaction when they take their morning walk. It makes them feel dreamy when they dress for dinner. One can predict that junk-science support for inherited privilege will reappear.

As for the eternal nature-nurture debate, one can offer the following caveat for all studies of the contributions of genes and environment to human intelligence: The genome can be important, but even more important is genetic expression, since it's expression of genes that determines which genes are active at any given moment in developmental history. What evidence exists suggests there is too much interplay between local environment and expression to justify assigning some fraction of the variance of a trait to "heredity" and some fraction to "environment." At the level of genes and molecules and cell biology, the dichotomy between"genes" and "environment" is based on mythology.

Similarly, the idea that "inherited" means "unchangeable" is a myth. For any trait partly governed by a set of genes, change the expression of those genes by environmental manipulation and the measure of the trait will change. What seems unchangeable today may be changeable tomorrow in ways unimaginable to us. Jensenism has unfortunately offered the public a fallacious reading of biological realities.

And meanwhile, as indicated in the numbers at the head of this chapter, Einstein's brain was indeed 11.3 percent smaller in weight than the average of age-matched controls, although some people, including Arthur Jensen, continue to tell us that brain weight predicts intelligence.

PART SEVEN

ON TRUTH
AND LIES

19

OUR FAILURES: IS ANYONE INNOCENT?

Science cannot stop while ethics catches up . . . and nobody
should expect scientists to do all the thinking for the country.
— Elvin Stackman

A public that does not understand how science works can all
too easily fall prey to those ignoramuses . . . who make fun of
what they do not understand, or to the sloganeers who pro-
claim scientists to be the mercenary warriors of today and the
tools of the military.

— Isaac Asimov (1920–1992)

WHAT ARE THE FORCES in modern society that foster junk science,
that make junk science possible, that make it a useful strategy? Who
benefits from twisted science? When junk science is successful,
meaning when junk science is accepted by the public, why does this happen?
Does it always happen? Why are some forms of twisted science readily accepted by the public and others brushed aside? Who is to blame for public
junk-science vulnerability and who is to blame for exploiting that vulnerability?

Research programs seeking the answers are in the province of sociology and
psychology, and not yet fully developed. But what seems clear about junk science in America is that one usually finds a conflict between special interest and
group interest, where special interest can be corporate or political or religionist
or simple class or racial or ethnic hatreds maybe derived from psychiatric dysfunctions, and group interest is public interest, the interest of the group at large,
the interest of all Americans.

What is also clear is that junk science can thrive in any free society where
the rule in the presentation of information to the public is caveat emptor—

buyer beware. If the "reality" offered is twisted, the public has no protection except its own devices and understanding, and an occasional assist from individuals and organizations who have only small voices in the public arena. And often the greater the twisting of reality the more easily the public is fooled. Many successful politicians understand this quite well. Adolf Hitler (1889–1945) understood this rule of corrupted reason: "The broad mass of a nation . . . will more easily fall victim to a big lie than to a small one" is his famous statement. Lying to the public is an old game, no doubt as old as human history, and a game with an apparently solid future in American life.

But admitting that the public is easily fooled is not an admission of defeat. There are thankfully enough people who understand that actions against the public interest ultimately demean their own existence, demean the existence of everyone, since beyond the cocoon of our own individuality what have we except our connections to others, to the larger group and to the human species? We may not be capable of eradicating the corruption of reason, but we must nevertheless counter it at every instance and with every means. In any large city, rats are a feature of the city's existence and cannot be completely eliminated. We understand that fact of life and we do our best to control the rat population so that it doesn't overwhelm us. What other course is there? Are we to abandon our cities to the rats?

THE FAILURES OF INDUSTRY

WHERE INDUSTRY AND JUNK SCIENCE are concerned, what is most striking from an examination of attitudes is the idea in many places that if social responsibility reduces profits, then actions that are socially responsible must be avoided or minimized. The general proposition is that a corporation exists to make profits for its shareholders, and that that objective must take precedence over any other consideration, even if society itself may be endangered by a proposed corporate action.

Where does this ridiculous notion come from? Which corrupted graduate business schools are turning out MBAs trained in such a fallacy? Who are the so-called thinkers who lead young business executives to the destructive fantasy that "hard-boiled" rapacious corporate-individualism is preferable to a strategy that considers group-interest an important adjunct to self-interest?

For more than three decades, the views of the economist Milton Friedman on corporate social responsibility have pervaded American business and the media. Friedman's ideas are known to every business school graduate and corporate executive, and the more conservative the business school, the less likely an MBA graduate of that school has ever heard serious counterarguments to Friedman's follies.

Friedman's major proposition on the question of corporate social responsibility was clearly stated in a magazine article in 1970:

> There is one and only one social responsibility of business — to use its resources and engage in activities designed to increase its profits so long as it stays within the rules of the game, which is to say, engages in open and free competition without deception or fraud.

The problem is not the ideas in the above quotation, but the repeated distortion of the ideas by both Friedman and his followers, the distortion accomplished by omitting the second half of the statement and avoiding any mention of "open and free competition without deception or fraud." Indeed, avoidance of deception and fraud is itself an exercise of social responsibility, any relevant laws merely a codification of that responsibility.

Friedman and his supporters usually avoid much discussion of the "rules of the game." The reason for the omission is that judgments about open competition, deception, or fraud must, in the real world, be made by parties other than the corporation, by the courts or by some government agency, either type of judgment often involving the enforcement or imposition of regulation of corporate behavior, an enforcement or imposition repugnant to laissez-faire economists and conservatives such as Friedman.

The problem is most obvious when it concerns corporations whose products or services are heavily dependent on science and technology, for example the chemical and pharmaceutical industries, since the science and technology behind the products and services in such industries is usually proprietary — secret and unknown to the public — and only the corporation knows when there is potential harm to the public. So what is the "social responsibility" of the corporation in such a case? Should the corporation continue a presently legal activity that could be potentially harmful to the public in the long term? Many Friedman free-marketers say yes, the corporation should continue the activity, because in the end, when damage to the public becomes obvious, the tort laws will make the activity unprofitable. Indeed, claims for damages may ultimately be a factor in controlling corporate behavior, but after how many deaths or injuries to innocent people? Can society afford unregulated corporate behavior when serious public harm, "collateral damage" in the struggle for profits, is a possible consequence? In such circumstances, can society afford the existence of corporate executives whose decisions are based solely on profits no matter what the consequences for society?

Friedman's 1970 denunciation of corporate social responsibility did serious damage to the American public, inciting tremendous anticorporate antagonisms that quickly translated into antiestablishment antagonisms—a popular revolt against a sophomoric analysis of the role of business in society by an economist with a following.

Have Friedman's ideas changed? Nearly thirty years after his 1970 analysis of corporate social responsibility, Friedman's views about laissez-faire capitalism remained as sophomoric as ever. Here is Friedman in a 1999 video interview explaining why child labor in Victorian laissez-faire London, the London of Charles Dickens, was not as bad as people believe:

> Now so far as child labor is concerned . . . what happens is, what happens in the picture that's drawn of Britain in the nineteenth century is that there's no image of what went before. Of why is it that all these people from the farming, from the rural areas came to the city. Did they come to the city because they thought it would be worse? Or because they thought it would be better? And was it worse or was it better? In the early days, you know there are very few things that are 100 percent black or 100 percent white, there are various shades of gray. And what we aim for is the least shade of gray that's possible. I'm not going to say that all was rosy in Britain at the time, it wasn't. But look around the world today. Where is it least rosy? In those countries where things that are run by the government not in those countries where private enterprises are. And the same thing was true in Britain, the conditions in the rural areas, on the farms, were far worse than conditions in the city, but they were not visible, they were hidden, nobody saw them.

In what academy would such an analysis pass as adequate? Apparently, Friedman's point is that nineteenth-century laissez-faire capitalism wasn't so bad in practice, since what existed before was worse, and what existed outside London had to be worse, he says, otherwise why else did people move to live in London? Of course, Friedman avoids the fact that at that time, in the nineteenth century, laissez-faire capitalism was in place both outside and inside London, and also in place everywhere in England prior to the nineteenth century—at least anywhere in England where capitalism existed. People gravitated to London because its population density provided opportunities for work, and not because the economic system in London was any better than elsewhere—the system was essentially the same everywhere in England—unregulated capitalism. More opportunities existed in cities because of population density, and not because of differences in economic systems between rural and urban areas.

Friedman can offer no defense of unregulated capitalism in nineteenth-century England. The stark moral question that Friedman avoids remains: Is a ten-year-old child slaving twelve hours a day in a textile mill a laudable achievement of the free market system? And was this abomination ended by the free market system itself or only by regulation, by the child labor laws?

As expected, in this interview as in many others, Friedman urges us to return to laissez-faire capitalism as the modus operandi of modern society. He favors abolishing the Food and Drug Administration, the Environmental Protection Agency, the Occupational Safety and Health Agency, and any other government agency that functions to protect the public against harm by corporate interests. We're to be "libertarians" in the wild, all of us free to do whatever we like in the jungle, free to live our individual lives without any connection to the group, without any connection to the human species, free to be isolated savages in the Friedman free-market jungle.

Child labor laws? In the Friedman free-market jungle there is no regulation of industry. Forget about child labor laws.

No thank you, Milton Friedman. Friedman's version of capitalism is unworthy and obsolete, made obsolete by the twentieth century unequivocally demonstrating that capitalism has more to offer humanity than child labor, debtor prisons, and maggots in meat. Evidently, some of our libertarian neo-conservative high-riding cowboys need to be reminded that the phrase "free society" has two words: The second word is "society."

Friedman's call for a return to laissez-faire capitalism is a product of wimpy thinking, his ideas an intellectual embarrassment. Unfortunately, as we have seen in this book, too many corporate executives use such ideas to justify socially destructive junk-science corporate behavior, and the consequent damage to the public interest has been significant.

THE FAILURES OF GOVERNMENT

INDUSTRY TWISTS SCIENCE TO MAINTAIN or increase profits, but the twisting of science by modern government is a different game. In a representative democracy like the United States, the corruption of science by government almost always occurs as an attempt to maintain or increase an electoral constituency. Otherwise, the corruption of science is usually an extension of industrial corruption of science with benefit of some kind to individual politicians or political parties. Of course, one can argue that in some cases government is as easily duped as the public itself, but to lump government and public together as victims may be unreasonable given the resources of government to protect itself against twisted science originating in the private sector. It seems to me the only approach that makes sense is to assume that when officials of government twist science, these officials know what they are doing—a tentative assumption until one has evidence to the contrary. Anyone in government—legislator, executive official, or member of the judiciary—has almost immediate access to every side of a science policy question if they want the information—a privilege unknown to members of the general public. So in the absence of contrary evidence, the most reasonable assumption is that when government corrupts science, government knows what it's doing.

The most prominent current example of the twisting of science by government is the ban on federally funded embryonic stem cell research, a ban rationalized by government-supported religionist thinking couched as "bioethics," the sum of the arguments junk science presented to the public as a basis for policy. But we can find many examples of modern government twisting science in various domains, including nutrition, environmental pollution, medical care, health care, support for antievolutionism, twisting of the facts concerning human cloning, global warming, missile defense, defense against terrorism, and so on. If industry is to be faulted for presenting junk science to the public, government is to be faulted even more so, since industry makes no pretense of public service, while public service is the ostensible reason for the existence of government in the first place. Because in the American system government exists by contract with the people, corruption of science by government is a profound and miserable travesty.

RELIGION AGAINST SCIENCE

IN RECENT YEARS, THE PERENNIAL conflict between science and religion has been discussed by some people as though the conflict were a mere disagreement resolvable by amicable discourse. This idea has been promoted in the media, on lecture tours, and in numerous books, the usual discussion concluding that these two arenas of human activity are intrinsically separate and each must be limited to its traditional domain—religion to matters of the spirit and morality, and science to matters of the real world—with each deserving of respect from the other. Such an understanding of the work of science and religion, for example, was the view of the evolutionist Stephen Jay Gould and is the current view of many religionist philosophers and essayists who broach the general question of how religion and science are to handle their historic conflict with each other.

The conflict between religion and science is of course relevant to any treatment of junk science, since were there no conflict there would be no twisting of science by religionists. Scientists, in their science, make no demands upon religion, do not use religion in their work, and have no reason to want religious beliefs to take one position or the other—except maybe any position that will leave science and the teaching of science alone. Religionists, in contrast, are deeply involved in science, since the discovery of the world by science continually contradicts many religious beliefs and has presented to religionists one philosophical burden after another through the ages. In short, religion may have an interest in twisting science, but science has no possible interest in twisting religion. That is the apparent problem that must be faced by any attempt to have religion and science holding hands and smiling at each other on the stage of human intellectual history.

But is there an even more fundamental problem? Maybe, in their noble attempts to make peace, the peacemakers have avoided the real problem. Four centuries ago, Francis Bacon (1561–1626) recognized an essential aspect of the division between philosophy (which in those days meant religion) and science: "Books must follow sciences, and not sciences books," Bacon said. The meaning is clear: If science discovers the real world, then philosophy and religion must deal with that real world as it is discovered. Absent that dealing, philosophy and religion sooner or later become obsolete and of no use to humanity. Many religionists recognize this; they refuse interpretations of religious texts that contradict current understanding of the real world as revealed by science;

they consider religion as an evolving human enterprise, evolving in parallel with science, unbound by doctrine that denies the realities of science. The world as we know it changes. It's not possible to argue that 50,000 or 100,000 years from now our understanding of the natural world will be the same as our understanding of the natural world today, and certainly many religio-philosophical ideas now extant among religionists and philosophers may become as obsolete as a literal reading of the Book of Genesis is now. What must change, science or religion? That is the fundamental question that I think is often ignored in discussions of the conflict between science and religion. As our species continues to discover the natural world, our spiritual response to that world must change accordingly. I see no other reasonable road for religion. And yes, this means all doctrine must eventually yield to realities.

But we're not there yet. During the past decade in America, we seem to be stepping backward, with some religionists advocating use of the Bible as a science text in public education. This cannot do, this will not happen in a free society, and it can never be more than a perversion of the human intellect.

POLITICS AGAINST SCIENCE

THE POLARIZATION OF AMERICAN POLITICS during the past decade has had a serious impact on the prevalence of junk science in public political discourse. Political debaters, standing on actual wooden platforms or writing in newspaper columns, are hardly ever evidentiary in their methods. For a political provocateur, the usual method is to be selective, to select evidence that supports the preferred agenda and ignore evidence against that agenda. Since the ill-informed public is almost always unaware of the evidence ignored, the political provocateur usually succeeds in making his case—until some provocateur on the other side presents contrary evidence, again carefully selected to support the opposing agenda. What has become common in America is the creation of "research" institutes that do no real research, but which publish polarized analysis or distorted data that can be cited as "evidence" by partisan hack newspaper columnists and other commentators.

For example, between 1989 and 1992 the conservative George C. Marshall Institute published a series of "scientific" papers endorsed by Frederick Seitz, the articles with anonymous authors, the analyses ostensibly supported by several reputable meteorologists, the reports insisting that action on the basis of feeble global warming theories would be extremely costly to the U.S. economy. As the physicist Spencer R. Weart points out,

> *Scientists noted something that the public largely overlooked: The most out-*
> *spoken scientific critiques of global warming predictions rarely appeared in*
> *the standard scientific publications. . . . With a few exceptions, the critiques*
> *tended to appear in venues funded by industrial groups and conservative*
> *foundations, or in business-oriented media like* The Wall Street Journal.

The problem is pervasive and serious and is nothing less than junk-science hucksterism. When the average American hears about a study published by a right-wing institute and he thinks it's a report of bona fide research by experts at a bona fide research institute, he has little understanding that his mind is being manipulated by political hucksters with a deep contempt for both the American public and a free society. And the political right is not alone in this game. Identical socially destructive behavior occurs on the political left, the public misinformed by junk science, misinformed by evidence twisted to suit some particular agenda, misinformed by arguments that science and scientists are socially "suspicious" entities whose evidence and conclusions are not to be trusted. The left, unfortunately, often sees science as merely a voice of the establishment. Indeed, one can argue that by focusing on science as a social product rather than on scientific discoveries of the real world, the left has abandoned the public and fomented its own suicide.

As a rule, the more polarized the political stance, the more likely realities are twisted to suit the arguments, and at the extremes of polarization it's no surprise to find that those on the extreme left and extreme right essentially practice the same methods in the use of junk science and propaganda. Their contempt for the American public is essentially identical, since both extreme right and extreme left believe the public must be led by a political elite—themselves. If you believe the public is too stupid to know its own interest, it's easy to also believe you can sell a political idea the way Madison Avenue sells soap. The use of junk science in the political arena is an exploitation of public ignorance by people with political agendas, people who devote much energy to publicly proclaiming their concern for the public, but who have an apparent contempt for the public, an insidious contempt dangerous to any free society. If and when Americans lose their freedom, they will be told, with contempt, that the loss is for their own good.

THE FAILURES OF THE MEDIA

DESPITE ITS FREQUENT MANIFESTOS OF professional responsibility, the media is not in public service, and most entities in the media do not exist to inform the public, but instead exist to entertain the public and sell advertising adjacent to or embedded in that entertainment. Even "nonprofit" media supported by the public have lost their way: These days, "nonprofit" merely means no shareholders rather than an absence of profiteering. "Nonprofit" profits are distributed internally as salaries and fringe benefits to managers rather than distributed externally to shareholders, a reality that usually escapes public attention.

Science has always been a problem for the media, since science is usually complex, its language usually jargonized, and people who choose careers in media often have little formal education in science and little informal knowledge of science fundamentals. The rule is that where science is involved media people are as easily fooled as the public at large. Were there true professional responsibility in the media, were there at least some trained advisors, purveyors of junk science would find it extremely difficult to use the media as their handmaidens. Instead, it's easy, much too easy, a game in which the media are easily manipulated because to most journalists and media executives science is a foreign country.

What the media understand is that people are interested in science even if they're not interested in learning any science. And of course the result is that science with entertainment value has priority over science related to important public issues. Consider the media from the standpoint of Milton Friedman's views of corporate social responsibility: To maximize profits, the media must focus on entertainment rather than on enlightenment, an idea internalized by every media executive. Of course, there are many audiences, and some audiences may find enlightenment entertaining, and so the media offer these small audiences the entertainment they want. But the media motive is hardly ever enlightenment as public service.

Some professionals in media and education believe themselves participants in a movement to increase public understanding of science. But too often the science of interest is too specialized even for other scientists to understand it, and understanding essentially impossible for the public and the media. What can be understood, however, is the most important aspect of the scientific attitude, the emphasis on hard objective evidence evaluated by hard objective reason. That emphasis can certainly be taught to every reasonably intelligent individual, but nothing like that happens because the focus of the media, and often the focus of educators, is popularized science presented as entertainment, with a mantra that all views must be considered equal, and a careful avoidance of any science that might be objectionable to special groups.

The notion that what should be presented to the public by the media is the idea that science is "fun" is both banal and a disservice to the public. Nuclear weapons are not "fun"; global warming is not "fun"; viral epidemics are not "fun"; tobacco poisoning is not "fun." The ignorance of television people who command high salaries to stroke the public ego is not funny at all. How depressing it is to watch television news broadcasters snickering or giggling about the complexity underlying some news story in science. Can one expect the news producers and news executives who condone this behavior to be at any higher level of intellectual achievement? Can we expect these people to be sophisticated enough to protect the public against junk science hawked in the interest of political or social or commercial agendas?

THE FAILURES OF AMERICAN EDUCATION

EVERYONE APPARENTLY AGREES THAT COMPARED to the rest of the world, American students lag in science achievement. Whatever the reasons for this, it's also true that education of American students at all levels has failed to produce an American public capable of protecting itself against exploitation by purveyors of junk science in the political, social, and economic arenas. Are people as easily exploited elsewhere? My guess is yes, and if they are, I propose the reason is the same: What is not taught to nonscience students is an attitude toward evidence that protects against exploitation.

The general idea in American science education through the eighth grade is that a large portion of school time should be devoted to exploring plants and animals, rocks and stars, electricity and gravity, and a host of other science topics, all presented in a way to sustain the interest of students. The belief is that the job of teachers between kindergarten and eighth grade is to "give children an understanding of the general laws of nature; make sure they know some important facts and ideas about the universe; train them in some methods scientists use; give them an appreciation of the scientific view of the world; and stimulate their desire to ask questions and find answers."

Those are worthwhile educational objectives, but such objectives will never produce a public protected against exploitation by junk science unless the emphasis every day in every science class is the importance of objective evidence. Children must be taught to ask the blunt question: "Where's the evidence?" They must ask it of their teachers, they must ask it of their parents, they must ask it of themselves. Until the goal of public education is to inculcate a philosophy of doubt and a critical faculty in children, the public will remain vulnerable to exploitation and grow more vulnerable as the gulf between scientific knowledge and public knowledge increases.

What I'm proposing here is that the emphasis in public education shift away from popularization of science in grade school to an emphasis on development of protective attitudes. In education in the lower grades, the popularization of science without a heavy emphasis on the nature and necessity of evidence is not in the public interest—it's a game of entertainment rather than a work of enlightenment. Children need to be taught about evidence and how to use it and how to think about it, and teaching such fundamentals can begin in early grades in ordinary language and continue through the end of high school. Most students will recognize the relevance to their daily lives of a focus on evidence, and they will be open to it. It's more important for a child to understand the nature of evidence than it is for the child to know that a frog is called an "amphibian."

The overwhelming majority of children will not become scientists. As adults, they will form the public addressed by both real science and by junk science, and a primary objective of our educational system must be to give them, by the time they finish high school, at least the ability to distinguish one from the other. If we fail at that, by the end of the twenty-first century, when the gulf between scientific realities and public knowledge is much wider, the exploitation of the public by junk science will have attained fearful proportions and the stability of society may be in danger, if it isn't already.

AT THE UNIVERSITY LEVEL, the problem of American education of nonscientists is of another kind, a serious problem of antiscience, of blatant antagonism toward science by many teachers and scholars in the humanities. These people believe that science is spiritually corrosive, that an understanding of optics, for example, strips away "the mysterious beauty of the rainbow," that knowledge of human internal anatomy and physiology somehow dehumanizes us. They refuse science, but it's a refusal laden with hypocrisy, since they enjoy the benefits of science and technology in their everyday lives, and when they or their children are ill they rush to a hospital and demand the latest in medical science as treatment. A pox on your science but do save me!

The two cultures articulated by C. P. Snow many years ago are still with us, more separate than ever, and the divide affects the education of every college student in America. The differences are so stark, the possibility of intellectual alliance seems hopeless. I think the most fundamental difference concerns where people look: Science looks to the future, to new knowledge, to new possibilities, while the humanities look to the past, to tradition, to literary and artistic golden ages, to aesthetic ideals of past centuries.

The attitude of many people in the humanities is illustrated by a public complaint by the novelist Fay Weldon to the scientific community: "We, the public, have to put up with your brave new world because there is no going back, and the past is an ignorant and brutal land and all of us were miserable there. But don't expect us to like you."

IF NOT TO BE liked is a sad thing for scientists, to be abused by silly deconstructionist ideas is worse. A cultural anthropologist provides her impressions of the science of immunology:

> As immunology describes it, bodies are imperiled nations continuously at war to quell alien invaders. These nations have sharply defined borders in space, which are constantly besieged and threatened. In their interiors there is great concern over the purity of the population—over who is a bona fide citizen and who may be carrying false papers. False intruders intend only destruction, and they are meted out only swift death. All this is written into "nature" at the level of the cell.

The author then wonders "whether this imagery might make analogous social practices come to seem even more natural, fundamentally rooted in reality, and unchangeable." And later the author adds: "Are there powerful links between the particular metaphors chosen to describe the body scientifically and features of our contemporary society that are related to gender, class, and race?"

Well, yes. But are the given metaphors "chosen," or simply natural language describing complex events? And is science to be condemned for its use of metaphor?

Immunology is one of the more difficult biological sciences, difficult in conceptualization and difficult in practice. It's essentially a new science, a twentieth-century science, and in terms of saving lives and reducing human suffering, immunology has paid its dues in full. Is the imagery used by immunologists really something to be condemned by colleagues across the quadrangle in the humanities building? If the imagery used by immunologists helps immunologists understand and teach their science, what is there to condemn?

Of course the real problem is that when students in a humanities lecture hall hear this sort of pontifical condemnatory analysis, or hear how the writer D.H. Lawrence condemned the science of optics for destroying "the mysterious beauty of the rainbow," they don't walk out of that lecture hall with a positive attitude toward science, and in a few years, these students join the public as people vulnerable to exploitation by junk science. Have the humanities faculties considered the public consequences of their antiscience attitudes?

Throughout human history, antiscience, whether popular or academic, whether promoted by tribal shamans or literary people or university professors, has been the mother of human exploitation. Can we assume that cultural anthropologists are aware of this as they write about the imagery of immunology?

THE FAILURES OF SCIENTISTS

AND HOW INNOCENT ARE THE scientists? Scientists don't do enough and they know it. They complain to each other that they do too little to combat the twisting of science, but if the reality of inaction is admitted, the situation remains hardly changed from one decade to the next. There's much irony. Most working scientists have little respect for nonscientist popularizers of science, but they also have little respect for popularizers who are scientists. There are many reasons for the objection to popularized science, and some reasons are reasonable. There are many fields in science, for example, that cannot be popularized without introducing distortion. The purest examples are in fields such as mathematical physics. How do you present the world as revealed by mathematical physics to a public that has no training in either physics or higher mathematics? It's not possible. Every physicist knows, for example, that the meat of Einstein's scientific papers cannot be found in his popularization for the public, or in anyone else's popularization of Einstein's work. A sense, maybe, but not the meat. The same is true for much of modern science. The public is given a sense, but not the meat, and telling the public that popularized science produces scientific "literacy" does the public a disservice.

Scientific "literacy" is not even widespread in science itself. Certainly, scientists understand their own "illiteracy" in fields other than their own. Stephen Jay Gould put it this way:

> We have now reached the point where most technical literature not only falls outside the possibility of public comprehension but also (as we all admit in honest moments) outside our own experience in scientific disciplines far removed from our personal expertise.

In general, complete public science literacy is not possible, and it's for that reason that scientists must come out of the laboratory as often as they can to combat the dangerous twisting of science that will only grow more rampant in this century. No one else but scientists can do it on a continuing basis, and certainly the scientific organizations need to rethink their public missions if they expect the thriving of science in a free society. The issue is not whether the public likes science enough to fund research; the issue is potential wreckage of society by special interests twisting science for their own ends. The battle against junk science must be fought by all people of reason, but particularly by scientists in the front lines, and wherever and whenever necessary. Avoiding this struggle may drive us all into darkness.

IN THE TWENTY-FIRST CENTURY, science will impact the public as never before, and twisted science will have its own destructive effects. The outlines are clear. Who is against science? Who will battle against science in this century, a century that will see humanity transformed as never before by science and technology? As always, the primary enemies of science will be any institution or group whose power is threatened by knowledge of the real world. Delineate threats to power and the battlefield is clarified.

Science is the human enterprise that allows us to control our destiny. To corrupt science is to corrupt reason, and to corrupt reason is to demolish our edge over the apes. Take that edge away and we don't have much, certainly not enough to justify the agonies of our existence.

Junk science is not merely another price we pay for a free society, it's an insidious rot that needs to be confronted wherever and whenever it arises. My belief is that if current patterns of twisting science continue, we face ultimate calamity. For if truth brings freedom, what can lies bring us but slavery?

SOURCES

Because this book is intended primarily for the general reader and is not a scholarly work, I have forgone the use of footnotes to identify references. The following notes indicate both direct sources and some suggested reading sources for those who wish to pursue various topics. For journal citations, the year, volume number, and first page of the article are given, in that order. Web sources are kept to a minimum because Web pages tend to vanish.

Introduction

The quotation by Albert Szent-Györgyi: *The Crazy Ape*, Philosophical Library (1970), p. 75. The comment by Jacques Barzun: Stephen Toulmin, *Foresight and Understanding*, foreword by Jacques Barzun, Harper & Row (1961), p. 9. An interesting discussion of Popper's views on science can be found in Steve Fuller, *Kuhn vs. Popper*, Columbia University Press (2004).

Chapter 1: Science, Junk Science, and Dogma

The quotation by St. Augustine of Hippo: *Sermons, Book 1*. The quotation by J. Robert Oppenheimer: *The Open Mind*, Simon & Schuster (1955), p. 114. The quotation by Galileo on the Jovian moons: Stillman Drake, *Discoveries and Opinions of Galileo*, Anchor Books (1957), p. 57. Also, Wade Rowland, *Galileo's Mistake*, Arcade Publishing (2001), p. 94. The indictment of Galileo: William R. Shea and Mariano Artigas, *Galileo in Rome*, Oxford University Press (2003), p. 193. A classical historical source concerning eighteenth- and nineteenth-century brain research is Edwin G. Boring, *A History of Experimental Psychology*, Appleton Century Crofts (1950). An example of a "genetic destiny" book: Aubrey Milunsky, *Your Genetic Destiny*, Perseus Publishing (2001). A short account of the eugenics movement in the early part of the twentieth century can be found in Garland E. Allen, *Science* (2001), 294:59. The quotation by Erwin Chargaff: *Perspectives in Biology and Medicine* (1973), 16: 492. A short account of eugenics in Nazi Germany: Susan Bachrach, *The New England Journal of Medicine* (2004), 351:5. Information about Ernst Ruedin can be found in *Nature* (2000), 407:823. Information about Lysenko can be found in Vadim J. Birstein, *The Perversion of Knowledge: The True Story of Soviet Science*, Westview Press (2001), p. 45. Details about Lamarckianism can be found in Monroe W. Strickberger, *Evolution*, second edition, Jones and Barlett (1996), p. 32. The quotation by Pavlov is given in Birstein, p. 16. Other sources: Giorgia de Santillana, *The Crime of Galileo*, University of Chicago Press (1955); for phrenology, Stanley Finger, *Origins of Neuroscience*, Oxford University Press (1994); Fred Watson, *Stargazer*, Da Capo Press (2004) (on Galileo's telescopes).

Chapter 2: Fraud and Fabrication in Science

The quotation by Kurt Cobain: *The Observer*, April 17, 1994. The quotation by Richard Lewontin: *The New York Review of Books*, November 18, 2004. Details about the Piltdown hoax: Jeffrey H. Schwartz, *Sudden Origins*, Wiley (1999), p. 107. For F. H. Edmonds on the dating of the Piltdown gravels: "Geology of the County near Lewes," *Memoirs of Geological Surveys* (1926) N. S., 319:68. For Stephen Jay Gould on the Piltdown hoax: *Hen's Teeth and Horse's Toes*, W. W. Norton & Company (1983), p. 201. Commentary by Richard E. Leakey and Roger Lewin, *Origins*, E. P. Dutton (1977), p. 33. History of enzyme nomenclature: Henry M. Leicester, *Development of Biochemical Concepts from Ancient to Modern Times*, Harvard University Press (1974), p. 176. On the Abderhalden case: U. Deichmann and B. Muller-Hill, *Nature* (1998), 393:109. On the Brach-Hermann-Mertelsmann affair: *Science* (2004), 303:1465; *Science* (2001), 291:1876; *Nature* (1998), 395:532. On the Erdmann-Caldwell quotation concerning the coelacanth fake: *Nature* (2000), 406:114; *Nature* (2000), 406:343. On the Schoen scandal in physics: *Nature* (2002), 420:728; *Nature* (2004), 429:789. On fraud in science: *Nature* (1999), 398:13. Other sources: Horace Freeland Judson, *The Great Betrayal: Fraud in Science*, Harcourt (2004); Robert L. Park, *Voodoo Science: The Road from Foolishness to Fraud*, Oxford University Press (2001); John E. Walsh, *Unraveling Piltdown: The Sci-*

ence Fraud of the Century and Its Solution, Random House (1996); Ronald Millar, The Piltdown Men, St. Martin's Press (1972); Daniel J. Kevles, "Fraud in Science," The Skeptic magazine, January 1, 1999; Ziauddin Sardar, "Practicing to Deceive," New Statesman, October 30, 1998; Robert Bell, Impure Science: Fraud, Compromise and Political Influence in Scientific Research, Wiley (1992); Albert M. Teich, "Impure Science: Fraud, Compromise, and Political Influence in Scientific Research," The Journal of Higher Education, September 1, 1993; Alexander Kohn, False Prophets: Fraud and Error in Science and Medicine, Blackwell (1988); David J. Miller and Michel Hersen (editors), Research Fraud in the Behavioral and Biomedical Sciences, Wiley (1992); William Broad and Nicholas Wade, Betrayers of the Truth, Touchstone Books (1983).

Chapter 3: The Food and Diet Circus: Eat Your Way to Happiness

The George Meredith quotation: The Ordeal of Richard Feverel (1859), Chapter 28. The Mary Evans Young quotation: The Observer, May 1, 1994. The reader's letter about eating: Chicago Tribune, February 3, 2005, section 1, p. 18. The letter from the chairman of the National Automatic Merchandising Association: Chicago Tribune, February 12, 2005, p. 26. The medical aspects of food and obesity: A. R. Shuldiner et al., The New England Journal of Medicine (2001), 345:1345. On obesity prevalence: D. M. Bravata et al., The Journal of the American Medical Association (JAMA) (2003), 289:1837. U.S. Government Dietary Guidelines and Michael Jacobson: Nature (2005), 433:186. The advertisement for pistachio nuts: Health (January/February 2005), p. 155. Calorie expenditure data: W. D. McArdle et al., Exercise Physiology, Lea & Febiger (1981); National Research Council, Recommended Dietary Allowances, tenth edition, National Academy of Sciences (1989). Excess weight in children: William H. Dietz, The New England Journal of Medicine (2004), 350:855. Comparative studies of weight-loss regimens: D. M. Bravata et al., JAMA (2003), 289:1837. On statins: D. J. Jenkins et al., JAMA (2003), 290:502. On nutritional genomics: The Scientist, January 17, 2005, p. 14. The article in Newsweek: January 17, 2005, p. 41. On Fen-Phen: The New York Times, February 12, 2005, p. B2. On Zantrex-3: Michael Specter, The New Yorker, February

2, 2004, p. 64. On vitamins: K. M. Fairfield and R. H. Fletcher, *JAMA* (2002), 287: 3116. On deregulation of the food supplement industry: *The New Yorker*, February 2, 2004, p. 64. On antibiotics in animal production: Stuart B. Levy, *The Antibiotic Paradox*, Perseus Publishing (2002), p. 149. On salmonella: D. G. White et al., *The New England Journal of Medicine* (2001), 345:1147. On the UK Swann Report: Stuart B. Levy, *The Antibiotic Paradox*. Perseus Publishing (2002), p. 152.

Chapter 4: GM Foods: Frankenstein in a Corn Patch

London Daily Mail headline: Daniel Charles, *Lords of the Harvest*. Perseus Publishing (2001), p. 237. Maurice Wilkins quotation: Horace F. Judson, *The Eighth Day of Creation*, Simon & Schuster (1979), p. 97. Jeremy Rifkin quotation: Jeremy Rifkin, *The Biotech Century*, Tarcher/Putnam (1999), p. 90. Martin Van Buren letter to Andrew Jackson: Adrian Berry, *The Next Ten Thousand Years*, New American Library (1974), p. 26 (no citation to original). The GM food scare in Europe in 1999: G. Gaskell et al., *Science* (1999), 285:384. On introducing new genes into plants: Stanton B. Gelvin, *Nature* (2005), 433:583. On modifications of the environment by organisms: P. M. Vitousek et al., *Science* (1997), 277:494. On modifications of the environment by humans: P. M. Vitousek et al., *Science* (1997), 277:494. On the Green Revolution: G. Conway and G. Toenniessen, *Nature* (1999), 402:C55. On critics of the Green Revolution: A. Smith, *Popline* (1992), 14:3. Paul F. Lurquin, *High Tech Harvest*, Westview Press (2002). On "yellow rice": Holger Breithaupt, *EMBO Reports* (2001), 2:1. On vitamin A deficiency: X. Ye et al., *Science* (2000), 287:303. On children at risk for vitamin A deficiency: Dean DellaPenna, *Science* (1999), 285:375. On genetic engineering of trace elements: B. Lonnerdal, *Journal of Nutrition* (May 2003), 133:1490S. On molecular breeding strategies: B. Mazur et al., *Science* (1999), 285:372. On GM food risk assessment: Paul F. Lurquin, *High Tech Harvest: Understanding Genetically Modified Food Plants*, Westview Press (2002), p. 178. On potential problems of genetically modified crops: A. Bakshi, *Journal of Toxicology and Environmental Health Part B: Critical Reviews* (2003), 6:211. Quotation by Richard Lewontin, *The New York Review of Books*, June 21, 2001, p. 81. Other sources: Desmond S. T. Nicholl, *An Introduction to Genetic Engineering*, Cambridge University Press (1994).

Chapter 5: Aging and Longevity:
Live Forever Now

The fountain of youth quotation is from www.hgh-pro.com. The Mary Lee Vance quotation: *The New England Journal of Medicine* (2003), 348:779. The Alex Comfort quotation: *Ageing: The Biology of Senescence*, Holt, Rinehart, and Winston (1964), second edition, p. 1. The quotation about convergence in the twenty-first century is from: James Hughes, *Citizen Cyborg*, Westview Press (2004), p. xii. The second quotation by James Hughes is from the same book, p. 23. For an example of neoconservative views of longevity see: Leon R. Kass, *Toward a More Natural Science*, Free Press (1985), p. 316. For details about hormones, see: E. E. Baulieu and P. A. Kelly, *Hormones*, Chapman and Hall (1990). The article by D. Rudman et al.: *The New England Journal of Medicine* (1990), 323:1. The journal editorial on HGH injections: J. M. Drazen, *The New England Journal of Medicine* (2003), 348:777. The commentary by Mary Lee Vance can be found in: *The New England Journal of Medicine* (2003), 348:779. The quotation concerning the antiaging benefits of meditation is from www.waystolongevity.com/Meditation. Aubrey Milunsky discusses the comparative life spans of animals in: *Your Genetic Destiny*, Perseus Publishing (2001), p. 270. Scott Gilbert discusses life expectancy in: *Developmental Biology*, sixth edition, Sinauer Association (2000), p. 574. The Rose and Long article: *Current Biology* (2002), 12:R311. A discussion of the genes involved in aging: David Gems, *Nature* (2001), 410:154. Information about cellular senescence: A. Krtolica et al., *Proceedings of the National Academy of Sciences* (2001), 98:12072. Comments and quotation by Leonard Hayflick: *Nature* (2000), 403:365. Comments about Barbara McClintock: S. A. Stewart and R. A. Weinberg, *Oncogene* (2002), 21:627. Information about telomerase: S. E. Artandi et al., *Proceedings of the National Academy of Sciences* (2002), 99:8191. Information about aging and DNA damage: J. de Boer et al., *Science* (2002), 296:1276. Other sources: Stephen S. Hall, *Merchants of Immortality*, Houghton Mifflin (2003).

Chapter 6: Tobacco: Drug Dealing in America

The quotation by Sandefur: *The New York Times*, September 7, 1997. The Feldman et al. quotation: *Principles of Neuropharmacology*, Sinauer Association (1997) p. 602. Statistics concerning tobacco deaths are given by: Nancy A. Rigotti, *The New England Journal of Medicine* (2001), 346:506. Estimates of future tobacco deaths are given in: Henry A. Waxman, *The New England Journal of Medicine* (2002), 346:936. An excellent reference about the various tobacco companies: Fred C. Pampel, *Tobacco Industry and Smoking*, Facts on File (2004), p. 48. On the first evidence for the link between cancer and cigarettes: E. Wynder and E. A. Graham, *JAMA* (1950), 143:329. The classic paper on the painting of cigarette tar on the backs of mice: E. Wynder et al., *Cancer Research* (1953), 13:855. The quotation by the UK Tobacco Institute, 1954: UK Tobacco Institute Research Council, "A Frank Statement to Cigarette Smokers," January 4, 1954. Quotation by R. J. Reynolds Tobacco Company, 1954: *Report of Special Master: Findings of Fact, Conclusions of Law and Recommendations Regarding Non-liggett Privilege Claims*, Minnesota Trial Court File Number C1-94-8565, March 8, 1998, quoting Pioneer Press, October 24, 1954. Statement by UK Imperial Tobacco, 1956: E. J. Partridge, letter to Sir John Hawton, March 9, 1956. The Addison Yeaman quotation 1963: C. Mollenkamp et al., *The People vs. Big Tobacco*, Bloomberg Press (1998), p. 250. The statement by a Philip Morris director, 1964: Howard Cullman, board member at Philip Morris, 1964, cited in R. Kluger, *Ashes to Ashes: America's Hundred-Year Cigarette War, the Public Health, and the Unabashed Triumph of Philip Morris*, Alfred A. Knopf (1996), p. 260. The quotation by the Philip Morris Tobacco Company, 1968: R. Kluger, p. 325, quoting Duns Review, April 1968. Quotation by Brown & Williamson, 1971: Brown & Williamson presentation, "The Smoking/Health Controversy: A View from the Other Side," February 8, 1971 (BW-W2-03113) (L&D BAT file 4) www.ash.org.uk/html/conduct/pdfs/advert.pdf. Quotation by UK Imperial Tobacco, 1975: Sir John Partridge, Chairman of Imperial Tobacco, answers questions put at the AGM by ASH, 1975. L&D Imp 23, www.ash.org.uk/html/conduct/pdfs/advert.pdf. Quotation by Philip Morris Tobacco Company, 1976: Thames Television, "Death in the West" (1976). Quotation by Tobacco Institute of Hong Kong, 1989: Tobacco Institute of Hong Kong Limited (Introducing the Tobacco Institute), March 1989, C.7., www.ash.org.uk/html/conduct/pdfs/advert.pdf. Quotation by UK Tobacco Institute, 1998: M. Walker, testimony at the Minnesota trial, 1998. Minnesota attorney general quotation by Geoffrey Bible, 1998: D. Shaffer, "No Proof That Smoking Causes Disease, Tobacco Chief Says," *Pioneer Press*, March 3, 1998. John Carlisle interview, 1998: Tobacco Manufacturers Association, *Punch*, April 11, 1998. The anecdote involving an actor and an R. J. Reynolds executive: Thames TV, "First Tuesday, Tobacco Wars," June 2, 1992.

Material on nicotine addiction: R. S. Feldman et al., *Principles of Neuropharmacology*, Sinauer Association (1997), p. 591. On the chemical contents of tobacco smoke and effects on the fetus: X. Wang et al., *JAMA* (2002), 287:195. Statistics concerning smokers in the United States: *CDC Morbidity and Mortality Weekly Report* (2000), 49:881. On passive smoking toxicity: R. Otsuka et al., *JAMA* (2001), 286:436. On nicotine yield of cigarettes: R. S. Feldman et al., p. 591. Nicotine as a pharmacological agent: R. S. Feldman et al., p. 597. Symptoms of nicotine withdrawal: R. S. Feldman et al., p. 604. Toxicity of nicotine: *Merck Manual of Diagnosis and Therapy*, seventeenth edition, Merck Research Laboratories (1999), p. 2643. On molecular interactions of nicotine and the nervous system: R. C. Hogg and D. Bertrand, *Science* (2004), 306:983. Quotation by tobacco executive James Morgan: *Time* magazine, May 12, 1997. The quotation by C. E. Teague: R. D. Hurt and C. R. Robertson, *JAMA* (1998), 280:1173. On tobacco industry efforts to alter the chemical form of nicotine: *JAMA* (1998), 280:1173. On tobacco industry targeting of young people: C. King et al., *The New England Journal of Medicine* (2001), 345:504. On trends in tobacco company expenditures: *The New England Journal of Medicine* (2001), 345:504. On tobacco industry targeting of young adults: P. M. Ling and S. A. Glantz: *JAMA* (2002), 287:2983. On psychological and social profiles of cigarette smokers: Steven A. Schroeder, *The New England Journal of Medicine* (2004), 350:293. On tobacco industry acquisitions: B. Shamasunder and L. Bero, *JAMA* (2002), 288:738. On tobacco and the U.S. government: Steven A. Schroeder, *The New England Journal of Medicine* (2004), 350:293. On package warnings in other countries: *The New England Journal of Medicine* (2004), 350:293.

Chapter 7: Junk Medicine, Big Pharma, Big Profits

The cocaine advertisement is from the National Library of Medicine and is reproduced in D. and E. M. Armstrong, *The Great American Medical Show*, Prentice Hall (1991), p. 161. The quotation by Marcia Angell: *The Truth About the Drug Companies*, Random House (2004), p. 153. Data concerning drug company expenditures: Cynthia Smith, *Health Affairs* (January–February 2004), p. 160; Marcia Angell, *The Truth About Drug Companies*, Random House (2004), p. xiv. Drug industry expenditures on physicians: *The New En-*

gland Journal of Medicine (2004) 351:1885. Article in the *British Medical Journal: British Medical Journal* (2003), 326:1193. Gifts to residents: *JAMA* (2000), 283:373; *Academic Emergency Medicine* (2004), 11:19. On the Kaiser Family Foundation Survey: "National Survey of Physicians Part 2," Kaiser Family Foundation (March 2002). On bribery of physicians: Marcia Angell, p. 126. The quotation by David Blumenthal: *The New England Journal of Medicine* (2004), 351:1885. On the Neurontin case: Marcia Angell, p. 161. The quotation by Marie Curie: Eve Curie, *Madame Curie*, Doubleday, Dan and Company (1938). *JAMA* articles concerning relations between universities and industry: E. A. Boyd and L. A. Berg: *JAMA* (2000), 284:2209; J. E. Bekelman et al., *JAMA* (2003), 289:454. On the Olivieri case: J. M. Drazen, *The New England Journal of Medicine* (2002) 347:1368. The quotation by Alexis de Tocqueville: *Democracy in America*, Oxford University Press (1946), p. 316. On the Bayh-Dole Act: Sheldon Krimsky, *Science in the Private Interest*, Rowman & Littlefield (2003), p. 165.

Chapter 8: Quack Doctoring: It's a Free Country, Isn't It?

The quotation by the nineteenth-century American doctor: A. E. Hertzler, *The Horse and Buggy Doctor*, Hertzler Research Foundation, New York (1938), p. 39. The quotation by James Lovelock: James Lovelock, *Science* (1998), 280:832. On medical schools in 1875: Otto L. Bettmann, *The Good Old Days—They Were Terrible!* Random House (1974), p. 142. On alternative medicine visits in the United States: D. M. Eisenberg et al., *JAMA* (1998), 280:1569. On the dangers of Ayurvedic herbals: Centers of Disease Control and Prevention, *Morbidity and Mortality Weekly Reports* (2004), 53:582–84; R. B. Saper et al., *JAMA* (2004), 292:2868. On the DSHE Act: L. M. Tierney, Jr., et al., *Current Medical Diagnosis and Treatment*, McGraw-Hill (2001), p. 1639. On ginkgo: C. A. Morris and J. Avorn, *JAMA* (2003), 290:1505. On St. John's wort: C. A. Morris and J. Avorn, *JAMA* (2003), 290:1505; Y. Y. He et al., *Photochemistry and Photobiology* (2004), 80:583; M. Martin-Facklam et al., *British Journal of Clinical Pharmacology* (2004), 58:437. On echinacea: R. B. Turner et al., *The New England Journal of Medicine* (2005), 353:341. On interactions between herbals and standard drugs: A. A. Izzo, *Fundamental and Clinical Pharmacology* (2005), 19:1. On echinacea: C. A. Morris and J. Avorn, *JAMA*

(2003), 290:1505. On herbal combinations and miscellaneous herbals: *JAMA* (2003), 290:1505. On sociological profiles of herbal users: John A. Astin, *JAMA* (1998), 279:1548. On arguments of advocates: P. B. Fontanarosa and G. D. Lundberg, *JAMA* (1998), 280:1618. On warnings by the Food and Drug Administration: S. E. Straus, *The New England Journal of Medicine* (2002), 347:2073. On toxic contents of herbals: S. E. Straus, *The New England Journal of Medicine* (2002), 347:2073. On adulteration of alternative medicines: S. E. Straus, *The New England Journal of Medicine* (2002), 347:2073. On chiropractic and McKenzie treatments: D. C. Cherkin et al., *The New England Journal of Medicine* (1998), 339:1021. On naturopathy: M. T. Murray and J. E. Pizzorno: *Encyclopedia of Natural Medicine*, second edition, Prima Publishing (1998), p. 13. On naturopathy treatment of glaucoma: M. T. Murray and J. E. Pizzorno, *Encyclopedia of Natural Medicine*, second edition, Prima Publishing (1998), p. 488. Other sources: D. Armstrong and E. M. Armstrong, *The Great American Medicine Show*, Prentice Hall (1991).

Chapter 9: Health Care in America: Sorry, Not for Everyone

The two White House quotations can be found at: www.whitehouse.gov/ infocus/technology/economic_policy200404/chap3.html. The quotation by President Johnson: *The Observer*, March 22, 1964. On U.S. life expectancies: Barry R. Bloom, *Nature* (1999), 402:C63. On the relation between economic class and longevity: S. L. Isaacs and S. A. Schroeder, *The New England Journal of Medicine* (2004), 351:1137; D. W. Baker et al., *The New England Journal of Medicine* (2001), 345:1106. On health care in rural America: S. J. Blumenthal and J. Kagen, *JAMA* (2002), 287:109. On cardiac catherization in women and blacks: K. A. Schulman et al. *The New England Journal of Medicine* (1999), 340:619. On disparities in the treatment of HIV patients: A. L. Gifford et al., *The New England Journal of Medicine* (2002), 346:1373. On neonatal mortality data: S. L. Lukacs and K. C. Schoendorf, *Morbidity and Mortality Weekly Report* (2004), 53:655. On mortality among non-Hispanic blacks: *Morbidity and Mortality Weekly Report* (2005), 54:1. On mortality among Hispanics: Centers for Disease Control and Prevention, *Morbidity and Mortality Weekly Report* (2004), 53:935. On the quality of health care in the United States: Barbara Starfield, *JAMA* (2000), 284:483. On medical errors: D. E. Altman et al.,

The New England Journal of Medicine (2004), 351:2041. On the problem of a health-care system based on market forces: Jerome A. Collins, *JAMA* (2000), 284:2184. On Canadian health expenditures: P. J. Veugelers and A. M. Yip, *Journal of Epidemiology and Community Health* (2003), 57:424. On the attitude of the Canadian medical community toward the American system: Stephen Bezruchka, *Canadian Medical Association Journal* (2001), 164:1701. On health and poverty: M. Marmot and R. G. Wilkinson, *Social Determinants of Health*, Oxford University Press (1999), p. 17; John P. Bunker, *International Journal of Epidemiology* (2001), 30:1260; the quotation by Michael Harrington is in: *The Other America*, MacMillan (1971), p. 202. Other sources: Geoffrey A. Rose, *The Strategy of Preventive Medicine*, Oxford University Press (1992).

Chapter 10: The Talk-Therapy Flea Market

The quotation by Thomas Szasz: *The Ethics of Psychoanalysis*, Basic Books (1974), pp. 161, 217. The quotation by Joel Kovel: *A Complete Guide to Therapy*, Pantheon Books (1976), p. xiii. On stress debriefing: Jerome Groopman, *The New Yorker*, January 26, 2004. On cognitive therapy and behavioral therapy: Michael A. Jenike, *The New England Journal of Medicine* (2004), 350:259. On modern psychiatry and the quotation by Hobson and Leonard: J. A. Hobson and J. A. Leonard, *Out of Its Mind: Psychiatry in Crisis*, Perseus Publishing (2001), p. 64. The quotation by E. Fuller Torrey: *The Death of Psychiatry*, Chilton/Haynes (1974). On the *Diagnostic and Statistical Manual of Mental Disorders*: Alix Spiegel, *The New Yorker*, January 3, 2005; American Psychiatric Association, DSM-IV, p. xxii; S. M. Turner and M. Hersen, *Adult Psychopathology and Diagnosis*, third edition, Wiley (1997), p. 3. Other sources: Arthur Burton (editor), *What Makes Behavior Change Possible?* Brunner/Mazel (1976). Other sources: T. M. Luhrmann, *Two Minds: The Growing Disorder in American Psychiatry*, Alfred A. Knopf (2000); Edward Shorter, *A History of Psychiatry: From the Era of the Asylum to the Age of Prozac*, Wiley (1997); R. L. Spitzer and D. F. Klein, *Evaluation of Psychological Therapies: Psychotherapies, Behavior Therapies, Drug Therapies, and Their Interactions*, the Johns Hopkins University Press (1970); E. Fuller Torrey, *The Death of Psychiatry*, Chilton Book Company (1974). E. H. Ackerknecht, *Short History of Psychiatry*, Hafner (1968); J. K. Wing et al., *The*

Measurement and Classification of Psychiatric Symptoms, Cambridge University Press (1974).

Chapter 11: Pollution: Private Interest and Public Poison

The Milton Friedman quotation is from an interview in *Playboy*, February 1973. The Warren Shaw quotation: *Fortune*, August 4, 1997. On tetraethyl lead: Herbert L. Needleman, *Environmental Research* (1997) 74:97; Devra Davis, *When Smoke Ran Like Water*, Basic Books (2002), pp. 65, 67; Jamie Lincoln Kitman, *The Nation*, March 20, 2000; R. L. Canfield et al., *The New England Journal of Medicine* (2003), 348:1517; D. A. Schaumberg et al., *JAMA* (2004) 292:2750. On air pollution: C. Arden Pope III, *The New England Journal of Medicine* (2004), 351:1132; J. M. Samet et al., *Science* (2004), 304:971; W. J. Gauderman et al., *The New England Journal of Medicine* (2004), 351:1057. On xenobiotics: R. J. Aitken et al., *Nature* (2004), 432:48. On mercury: J. G. Hardman et al., *The Pharmacological Basis of Therapeutics*, McGraw-Hill (1996), p. 1655; S. E. Schober et al., *JAMA* (2003), 289:1667. On women and organic solvents: S. Khattak et al., *JAMA* (1999), 281:1106. On high-voltage transmission lines: Linet et al., *The New England Journal of Medicine* (1997), 337:1; *Science* (1999), 285:23. The EPA quotation can be found at their Web site: www.epa.gov.

Chapter 12: Missiles and Terrorism: No Place to Hide

The John Derbyshire quotation: *National Review*, July 4, 2005, p. 48. The Martin Rees quotation: Martin Rees, *Our Final Hour*, Basic Books (2003), p. 44. On nuclear warheads: K. O'Nions et al., *Nature* (2002), 415:853. On rogue state nuclear strategy: Steven Weinberg, *The New York Review of Books*, February 14, 2002. On physicists and missile defense: D. Kleppner et al., *Physics Today* (January 2004); C. Eisendrath et al., *The Phantom Defense*, Praeger (2001), p. 11. On terrorism: C. Arnst and W. C. Symonds, *BusinessWeek*, October 1, 2001; Vadim J. Birstein, *The Perversion of Knowledge: The True Story of Soviet*

Science, Westview Press (2001), p. 303; F. A. Mettler et al., *The New England Journal of Medicine* (2002), 346:1554. On the nerve gas sarin: Ernest C. Lee, *JAMA* (2003), 290:659; J. P. Fitch et al., *Science* (2003), 302:1350. On terrorist chemical weapons: Lois R. Ember, *Chemical and Engineering News*, September 10, 2001. On biological terrorism: T. C. Dixon et al., *The New England Journal of Medicine* (1999), 341:815; T. V. Inglesby et al., *JAMA* (2002), 287:2236.

Chapter 13: Global Warming:
Yes, Your Beach House May Be Gone

The quotation by Senator Inhofe: *Bloomberg News*, January 21, 2005. On the IPCC: Bill McKibben, *The New York Review of Books*, July 5, 2001; report of the Intergovernmental Panel of Climate Change, 2001, Working Group 1: the scientific basis. The quotations by Frederick Seitz are from www.heartland.org/Article.cfm?artId=812. Views of other scientific groups on global warming: National Academy of Sciences Committee on the Science of Climate Change, *Climate Change Science*, National Academy Press (2001), p. 3; American Meteorological Society, *Bulletin of the American Meteorology Society* (2003), 84:508; American Geophysical Union, *EOS* (2003), 84:574; *American Association for the Advancement of Science*, www.ourplanet/aaas/pages/atmos02.html. On natural climate variability: T. J. Osborn and K. R. Briffa, *Science* (2004), 306:621. On continental glaciation: T. J. Crowley and R. A. Berner, *Science* (2001), 292:870. On climate during the last few million years: R. B. Alley et al., *Proceedings of the National Academy of Sciences* (1999), 96:9987. On demonstrations of scientific proof about the future: report of the Intergovernmental Panel of Climate Change, 2001, Working Group 1: the scientific basis, p. 13. On expected impacts: report of the Intergovernmental Panel of Climate Change 2001, Working Group 2: impacts, adaptation, and vulnerability, pp. 5, 6. On climate models: Richard John Huggett, *Environmental Change*, Routledge (1997), p. 27. On the recent history of climatology: Spencer R. Weart, *The Discovery of Global Warming*, Harvard University Press (2003), p. 187. Other sources: John Houghton, *Global Warming*, Cambridge University Press (1994); S. Fred Singer, *Hot Talk Cold Science*, the Independent Institute (1997); A. Upgren and J. Stock, *Weather*, Perseus Publishing (2000); Patrick J. Michaels, *Meltdown*, Cato Institute (2004).

Chapter 14: Creationism: The World as an Egg

The quotation by Pat Buchanan: *San Francisco Chronicle*, November 27, 1995. The quotation by Richard Dawkins: *The Blind Watchmaker*, W.W. Norton (1996), p. 316. The quotation by Alexis de Tocqueville: *Democracy in America*, Oxford University Press (1946), p. 240. The quotation by Pope John Paul II: *The Quarterly Review of Biology* (1997), 72:382. The declaration by Cardinal Schoenbron: *The New York Times*, July 7, 2005, p. A27. Evidence against a young Earth: Michael A. Seeds, *Horizons*, Wadsworth Publishing (1995), p. 356. Biblical literalism is certainly not mainstream Christianity and was in fact already denounced by St. Augustine in the fifth century A.D. in *The Literal Meaning of Genesis, Book I*. See H. Allen Orr, *The New York Review of Books*, February 26, 2004. On non-fundamentalist creationism and the quotations by John F. Haught: *Deeper Than Darwin*, Westview Press (2003); pp. 86, 118. On the OpEd essay by Michael J. Behe: *The New York Times*, February 7, 2005, p. A27. On the various vitalisms: Eric Nordenskiold, *The History of Biology*, Alfred A. Knopf (1935), p. 603. The AAAS resolution against intelligent design theory can be found at AAAS Board Resolution, October 18, 2002, www.aaas.org/news/releases/2002/1106id2.shtml. On the Dover, PA, evolution case: *Science* (2005), 307:505; with quotations from Dover School District, December 14, 2004: www.dover.k12.pa.us/doversd/cwp/view.asp?A=3&Q=261852. On the film "Volcanoes" and other films: *The New York Times*, March 19, 2005, p. A1. On intimidation of school teachers: *The New York Times*, February 1, 2005, p. D1. Other sources: Robert T. Pennock, *Tower of Babel*, MIT Press (1999).

Chapter 15: Stem Cells: The Petri Dish Blues

The quotation by Leon Kass: "The Wisdom of Repugnance" in L. R. Kass and J. Q. Wilson, *The Ethics of Human Cloning*, AEI Press (1998). The Christian P. Erickson quotation: *The New England Journal of Medicine* (2004), 351:1688. On human embryonic stem cells: Curt R. Freed, *Proceedings of the National Academy of Sciences* (2002), 99:1755, 2344; E. H. Kaji and J. M. Leiden, *JAMA* (2001), 285:545; Robin Lovell-Badge, *Nature* (2001), 414:88. The quotations by Charles Krauthammer are in: *The Weekly Standard*, August 20, 2001. On stem cell research since August 2001: George Q. Daley, *The New*

England Journal of Medicine (2004), 351:627. On Francis Fukuyama and Eric Cohen: Brian Alexander, *Rapture,* Basic Books (2003), p. 129. Other sources: Ronald M. Green, *The Human Embryo Research Debates: Bioethics in the Vortex of Controversy,* Oxford University Press (2001); Victor Hamburger, *The Heritage of Experimental Embryology: Hans Spemann and the Organizer,* Oxford University Press (1988); T. J. Horder et al., *A History of Embryology,* Cambridge University Press (1986); Albert R. Jonsen, *The Birth of Bioethics,* Oxford University Press (1998); National Institutes of Health, *Stem Cells: Scientific Progress and Future Research Directions,* National Institutes of Health (2001).

Chapter 16: Cloning: Will We Have a Hundred Jerry Falwells?

The quotation by Alvin Toffler: *Future Shock,* Random House (1970), p. 9. The quotation by Ian Wilmut: H. Gardner et al., *Good Work,* Basic Books (2001), p. 115. The quotation from the editorial in *The New York Times:* September 12, 1963. A comment on Toffler's impact: Lee M. Silver, *Nature* (2001), 412:21. On cloning techniques: J. B. Gurdon and A. Coleman, *Nature* (1999), 402:743; L. J. Kleinsmith and V. M. Kish, *Principles of Cell and Molecular Biology,* HarperCollins (1995), p. 445; Scott F. Gilbert, *Developmental Biology,* sixth edition, Sinauer Associates (2000), p. 85. On the cloning of a dog: *The New York Times,* August 4, 2005, p. 1. On the slippery-slope argument against cloning and various quotations by members of the President's Council on Bioethics: http://bioethics.gov. On limitations of reproductive cloning of humans: Rudolf Jaenisch, *The New England Journal of Medicine* (2004), 351:2787; I. Wilmut et al., *Nature* (2002), 419:583; R. P. Lanza et al., *Science* (2001), 294:1893. The quotation by Stephen S. Hall is from: Stephen S. Hall, *Merchants of Immortality,* Houghton Mifflin (2003), p. 354. On the cloning moratorium: *Science* (2005), 307:1702. Other sources: Lori B. Andrews, *The Clone Age: Adventures in the New World of Reproductive Technology,* Henry Holt (1999); Gina Kolata, *Clone: The Road to Dolly and the Path Ahead,* William Morrow (1998); Lee M. Silver, *Remaking Eden: How Genetic Engineering and Cloning Will Transform the American Family,* Avon (1997); Ian Wilmut et al., *The Second Creation: Dolly and the Age of Biological Control,* Farrar, Straus and Giroux (2000).

Chapter 17: Genes and Behavior:
You Need a Golf Gene Here

The quotation by John Alcock: *The Triumph of Sociobiology*, Oxford University Press (2001), p. 38. The quotation by H. Allen Orr: *The New York Review of Books*, February 27, 2003. Wilson's new synthesis: E. O. Wilson, *Sociobiology: The New Synthesis*, Harvard University Press (1975). The quotation by Tinbergen: N. Tinbergen, *The Animal in Its World*, George Allen & Unwin (1972). On proximate and ultimate causes in sociobiology: John Alcock, p. 12. The quotation by Philip Kitcher is from: *Vaulting Ambition*, MIT Press (1985), p. 435. On evolutionary psychology: John Alcock, p. 189; Stephen Jay Gould, *Hen's Teeth and Horse's Toes*, W. W. Norton (1983), p. 243. On the XYY chromosome fiasco: P. A. Jacobs et al., *Nature* (1965), 208:1351; L. F. Jarvik et al., *American Psychologist* (1973), 28:674; H. A. Witkin et al., *Science* (1976), 193:4253; M. J. Gotz et al., *Psychological Medicine* (1999), 29:953. On Einstein's brain: *The New York Times*, June 24, 1999; S. F. Witelson et al., *Lancet* (1999), 353:2149; Steven Pinker, *The Blank Slate*, Viking (2002), p. 44. On genetic testing: Neil A. Holtzman, *Science* (1999), 286:409; *The Merck Manual of Diagnosis and Therapy*, seventeenth edition, Merck Research Laboratories (1999), p. 2471; *Nature* (1998), 395:309.

Chapter 18: Race and IQ:
Two Myths Make a Bumble

The quotation by Arthur R. Jensen: Frank Miele, *Intelligence, Race, and Genetics*, Westview Press (2002), p. 136. The autopsy results for Einstein's brain: S. F. Witelson et al., *Lancet* (1999), 353:2149. On the Scholastic Aptitude Test: Joseph L. Graves, Jr., *The Emperor's New Clothes*, Rutgers University Press (2001), p. 158. Jensen's 1969 Harvard Review paper: A. R. Jensen, *Harvard Educational Review* (1969), 39:1. On the g factor: Frank Miele, p. 51; Arthur R. Jensen, *Bias in Mental Testing*, Free Press (1979), p. 251; Edwin G. Boring, *A History of Experimental Psychology*, second edition, Appleton-Century-Crofts (1950), p. 481; Frank Miele, p. 55; Stephen Jay Gould, *The Mismeasure of Man*, W. W. Norton (1996), p. 48. On Jensen's views on heredity and intelligence: Frank Miele, pp. 94, 96. On fetal environment: S. L. Lukacs and K. C.

Schoendorf, *Morbidity and Mortality Weekly Report* (2004), 53:655; L. Miller et al., *Morbidity and Mortality Weekly Report* (2002), 51:433; R. A. de la Chica et al., *JAMA* (2005), 293:1212. On current views of American psychologists concerning race: R. S. Cooper, *American Psychology* (2005), 60:60. On the history of race in anthropology: D. A. Hughes et al., *Current Biology* (2004), 14:R367; Monroe W. Strickberger, *Evolution*, second edition, Jones and Bartlett (1996), p. 548. On Jensen and the Pioneer Fund: Frank Miele, pp. 148, 154. On the failure of Jensenism: Frank Miele, pp. 88, 102. On other views about heredity and intelligence: Richard J. Herrnstein and Charles Murray, *The Bell Curve*, Free Press (1994), p. 122; 1995 afterword by C. Murray, pp. 526, 567. The quotation by Sarich and Miele: Vincent Sarich and Frank Miele, *Race*, Westview Press (2004), p. 194. On critiques of intelligence testing: Howard Gardner, *Frames of Mind*, Basic Books (2004), p. 16; Stephen Jay Gould, *The Mismeasure of Man*, W. W. Norton (1996). Other sources: J. P. Guilford, *Psychometric Methods*, McGraw-Hill (1936) (on factor analysis); Warwick Anderson, *The Cultivation of Whiteness*, Basic Books (2003) (on racism in Australia).

Chapter 19: Our Failures: Is Anyone Innocent?

The quotation by Elvin Stackman: *LIFE Magazine*, January 9, 1950. The quotation by Isaac Asimov is in the introduction to: Lewis Wolpert, *The Unnatural Nature of Science*, Harvard University Press (1993), p. ix. The quotation by Hitler: Adolf Hitler, *My Struggle*, chapter 10. Quotations by Milton Friedman: *The New York Times Magazine*, September 13, 1970; Hoover Institute film, February 10, 1999. On the George C. Marshall Institute and global warming: Spencer R. Weart, *The Discovery of Global Warming*, Harvard University Press (2003), p. 166. The quotation on the job of teachers in science education: W. J. Bennett et al., *The Educated Child*, the Free Press (1999), p. 345. The Fay Weldon quotation is from: *The Daily Telegraph*, December 2, 1991, p. 14. On the quotation by a cultural anthropologist: Emily Martin, *Medical Anthropology Quarterly* (1990) vol. 4, in Mario Biagiolo, *The Science Studies Reader*, Routledge (1999), p. 358. The quotation by Stephen Jay Gould on scientific literacy is from: *Science* (1999), 286:899. Other sources: H. L. Nieburg, *In the Name of Science*, Quadrangle Books (1966) (on science and government); Daniel S.

Greenberg, *Science, Money, and Politics,* the University of Chicago Press (2001); J. Gregory and S. Miller, *Science in Public: Communication, Culture, and Credibility,* Plenum Press (1998); Chris Mooney, *The Republican War on Science,* Basic Books (2005).

INDEX